S0-BKW-225

Land Treatment and Disposal of Municipal and Industrial Wastewater

Land Treatment and Disposal of Municipal and Industrial Wastewater

edited by

Robert L. Sanks
Professor
Department of Civil Engineering and Engineering Mechanics
Montana State University
Bozeman, Montana

Takashi Asano
Associate Professor
Department of Civil and Environmental Engineering
Washington State University
Pullman, Washington

ANN ARBOR SCIENCE
PUBLISHERS INC
P.O. BOX 1425 • ANN ARBOR, MICH. 48106

Second Printing, 1976

P. O. Box 1425, Ann Arbor, Michigan 48106

Library of Congress Catalog Card No. 75-10417
ISBN 0-250-40105-3

Manufactured in the United States of America

PREFACE

Land treatment and disposal, the practice of applying municipal and industrial wastewaters, treated effluents, and sludges to land, has received considerable attention for more than a decade and, within the last four or five years, much publicity. Previously, the objective was simply the disposal of wastes, but more recently, an increasingly important goal has been the utilization of nutrients in the wastes for raising useful crops and tertiary (or advanced) treatment of the water for improving stream quality or to recharge falling water tables. The need for conserving our nation's water resources and the requirements of the Federal Water Pollution Control Act Amendments of 1972 make it both desirable and necessary to investigate land treatment as one of the available alternatives in waste management systems.

The purpose of this book is to aid engineers and scientists engaged in investigating, planning, and designing systems for land treatment and disposal of wastewaters. The book is an outgrowth of the proceedings of the *Fourth Environmental Engineers' Conference* held at Montana State University. All the original chapters that were retained, however, have been revised, and more than a third of the chapters are new. All chapters reflect the latest technology updated at least to March, 1975. The first chapter deals with the regulatory agency viewpoint. Chapters 2 through 6 provide essential background information. Chapters 7 and 8 treat specific kinds of wastes, and Chapter 9 sounds a warning that all should heed. Chapter 10 deals with high rate systems, while Chapters 11 and 12 treat investigation, design and operation of land application facilities. The lessons of Muskegon, as far as they are known to date, are covered in Chapter 13. Overland runoff is not included because the technology is developing so rapidly that current information would soon be obsolete.

The editors wish to express their grateful appreciation to the contributing authors for their cooperation.

Robert L. Sanks, Bozeman, Montana
Takashi Asano, Pullman, Washington
August 31, 1975

ABBREVIATIONS AND CONVERSIONS

Symbol	Description	Multiplier	Product Symbol
ac	Acre	0.4047	ha
avg	Average		
bhp	Brake horsepower	0.7457	kW
cal	Calories	0.003968	Btu
cfs	Cubic feet per second	0.02832	m^3/sec
cm	Centimeters	0.3937	in.
cu ft	Cubic feet	0.02832	m^3
ft	Feet	0.3048	m
fps	Feet per second	0.3048	m/sec
g	Grams	0.00220	lb
gal	Gallons	0.003785	m^3
gal/ac	Gallons per acre	0.00935	m^3/ha
gpad	Gallons per acre per day	0.00935	m^3/ha day
gpcd	Gallons per capita per day		
gpd	Gallons per day		
gpd/sq ft	Gallons per day per square foot	0.283	m^3/min ha
gpm	Gallons per minute	0.0631	l/sec
ha	Hectares	2.471	ac
hp	Horsepower	0.7457	kW
hr	Hours		
in.	Inches	2.54	cm
kg	Kilograms	2.204	lb
km/hr	Kilometers per hour	0.6215	mph
l	Liters	0.2642	gal
lb	Pounds	453.59	gm
lb/ac wk	Pounds per acre per week	1.12	kg/ha wk
lb BOD/ac day	Pounds BOD per acre per day	0.89	kg/ha day
me	Milliequivalents		
mg	Milligrams		
mgd	Million gallons per day	0.04381	m^3/sec
mg/l	Milligrams/liter	1.00	ppm
mi	Miles	1.609	km
min	Minutes		
ml	Milliliters		
mph	Miles per hour	1.609	km/hr
ppm	Parts per million	1.00	mg/l
psi	Pounds per square inch	703.1	kgf/m^2
psig	Pounds per square inch gage	0.0703	kg/cm^2
sec	Seconds		
sq ft	Square feet	0.09290	m^2
T/ac	Tons per acre	2.24	t/ha
TDS	Total dissolved solids		
t/ha	Metric tons per hectare	0.446	T/ac
vol	Volume		
wk	Weeks		
yr	Years		
μg	Microgram	10^6	g

CONTENTS

1

LAND TREATMENT OF WASTEWATERS INSTITUTIONAL AND REGULATORY AGENCY APPROACH

Donald G. Willems

Chief, Water Quality Bureau
Environmental Sciences Division
Montana Department of Health and
Environmental Sciences
Helena, Montana 59601

An assessment of past wastewater treatment practices, existing and proposed laws, and regulatory standards can aid us in satisfactorily progressing toward higher wastewater treatment goals that are being required by state and federal laws. Montana's water pollution control program will be used as an example because it is probably typical of many states, at least of the Great Plains and Rocky Mountain areas. Public Law 92-500 (Federal Water Pollution Control Act Amendments of 1972) and regulations issued by the Federal Environmental Protection Agency (EPA) pursuant to it have much bearing on the acceptance of land disposal as a treatment alternative and will also be discussed.

PAST TREATMENT PRACTICES

Municipal Sewage Treatment

Montana's first treatment facilities were constructed in the early 1900s following passage of legislation for protection of domestic water supplies. The treatment facilities consisted of septic or Imhoff tanks with some being followed by subsurface disposal or sand filters. Most were built by small communities constructing new sewerage systems. For the most part the subsurface disposal systems and sand filters failed at an early date, leaving only the settling chambers to provide treatment. These failures can be related to inadequate design and construction and lack of good operation and maintenance practices.

The state's largest cities began construction of mechanical primary treatment facilities in the late 1940s. These plants improved conditions in the immediate area below the discharge but did little to improve stream conditions downstream. Most were constructed on a minimal budget and lacked sufficient flexibility. By-passing of facilities generally occurred at some time during the year.

The first biological secondary sewage treatment facility was constructed in Montana in 1948, and several other small communities also selected this means of treatment. These were of good design, but usually were poorly operated and maintained. As a comparison, two plants in the state of essentially the same design and providing treatment to similar flows provide 95% and 40% biochemical oxygen demand (BOD) removal, respectively. The first has a good operator and the second does not.

In the early 1950s, the first sewage stabilization pond was constructed in Montana. In the next 20 years, almost all of the state's communities with populations of less than 5000 and having sewerage systems constructed sewage stabilization ponds. These were readily accepted because of their relatively low capital costs and minimal operation and maintenance expense. Overall, they have provided a treatment between primary and secondary. In some cases they have created an odor nuisance during a short part of the year.

Land disposal of wastewaters in Montana is not new. For example, near the city of Helena, a farmer has used sewage effluent for irrigation about five months of each year since the early 1900s. The site is removed from residences, and there have been no complaints of the use of sewage for irrigation. However, complaints have been received from downstream residents when the farmer has diverted sewage effluent back to the receiving stream during the warm weather months as obnoxious odors occurred. No groundwater monitoring has been provided.

There are other examples of sewage effluents being used for irrigation in Montana. But again, there has been no groundwater monitoring. All the earlier systems were not designed but merely happened because of the convenience of irrigation water to the adjacent land owner. Systems are now being designed and constructed primarily as a means of treatment. As an example, a large resort complex under construction will use irrigation as final treatment for its wastewater. This means of treatment is being utilized to meet Section 69-4808.2(1)(c)(iii) of the Revised Codes of Montana, which states:

> The board shall require that any state waters whose existing quality is better than the established standards as of the date on which the standards become effective be maintained at that high quality unless it has been affirmatively demonstrated to the board that a change is justifiable as a result of necessary economic or social development and will not preclude present and anticipated use of these waters.

Industrial Wastewater Treatment

Industrial waste treatment has proceeded at about the same pace except that little was done by industries until passage of the Montana State Water Pollution Control Act in 1955.

Industrial wastewaters have also received additional treatment by application to the land. A pulp and paper mill in the state has disposed of a major portion of its effluent to the groundwater through pond seepage since 1958. Treatment efficiency is monitored through test wells, although the monitoring system is inadequate to assess fully the potential for groundwater damage off the mill's property. This means of disposal has provided good protection of the nearby stream. A meat packing company diverts stabilization pond effluent over a hillside. When the distribution system has been properly maintained, the overland flow has provided a highly improved effluent.

Summary of Past Treatment

Past treatment practices can, in general, be summarized as follows:

1. Inadequate design, construction, and flexibility are due mostly to insufficient funding.
2. Poor operation and maintenance are the rule, particularly in smaller communities.
3. Smaller communities accepted fairly reliable sewage stabilization ponds because of low operation and maintenance costs.
4. Poor land disposal practices were used accompanied by lack of monitoring and design.

From the assessment of past practices, the following factors in design must be recognized:

1. If effluent limits are to be met continuously, the necessary money to provide for adequate design, construction, and flexibility of treatment facilities must be provided.
2. A treatment system with low operation and maintenance is a must for successful treatment results from a smaller community.
3. Adequate monitoring must be included as part of the capital improvement and operation and maintenance budget to insure that the needed treatment is being provided.

PUBLIC LAW 92-500

Section 201

Congress, through Public Law 92-500, attempted on a nationwide basis to overcome some of the prior deficiencies in waste treatment management. Section 201 of the act (facilities planning for construction grant projects) has a stated purpose: "to require and to assist the development and implementation of waste treatment management plans and practices which will achieve the goals of the act." It requires study of alternative waste management techniques (including land application techniques) for the application of the best practicable waste treatment technology (BPT). It encourages recycling of wastewaters and treatment system management that combine open space and recreational considerations.

Section 202

Section 202 provides 75% grants to public bodies for construction of sewage treatment facilities. This should aid in overcoming the past practice of construction of facilities that barely, if at all, meet minimum requirements, which is typical when the majority of funds for construction are to be produced by the local government.

Sections 301 and 302

Section 301 requires the implementation of BPT by all point source discharges, except municipal discharges, by July 1, 1977. Municipal discharges are required to have secondary treatment by July 1, 1977, and BPT by July 1, 1983. All dischargers, except municipal, are required to meet best available technology economically achievable (BAT) by July 1, 1983. Both BPT and BAT have been or will be defined by EPA by category or class. For some categories, BAT may be defined by EPA as no discharge of pollutants. Section 302 further requires that additional effluent limitations be placed on discharges to streams where category effluent limitations are not great enough to meet water quality standards.

Section 402

Section 402 establishes a national waste discharge permit program. Through the permit, effluent limits are established for point source dischargers to surface waters and self-monitoring is required for both industrial and domestic sewage discharges. Substantial fines can be sought for permit violations. Through this section and enforcement provisions, the importance of providing a design with an adequate safety factor to meet effluent limits continuously becomes apparent.

Goal

One of the national goals in Public Law 92-500 is to eliminate the discharge of pollutants into navigable waters by 1985. One of the alternatives which, in some instances, appears to meet this federal goal is treatment by land application.

The Muskegon, Michigan, sewage irrigation project received much publicity at the time Public Law 92-500 was developed (see Chapter 13) and undoubtedly had substantial influence on some of its wording. This project has not been without problems. Projects such as this need a thorough evaluation to aid designers and regulatory agencies in reaching their goal of eliminating the discharge of pollutants to navigable waters.

REGULATIONS AND GUIDELINES FOR LAND TREATMENT

Present Status

Most states have not provided detailed regulations or guidelines for land disposal practices, and most regulations published have dealt with the treatment needed before effluents are applied to a particular crop or to land for a specific usage. Following passage of Public Law 92-500, it appeared that EPA would require secondary treatment and chlorination for all sewage before it was applied to a land treatment system. This approach may be practicable where human contact is made with the sewage immediately after application, but if there is little chance of human contact or if crops raised on the land are not eaten raw, a lower degree of treatment would appear applicable because in general the soils will provide a high degree of treatment. EPA now appears to be flexible on pretreatment requirements.

Both the designer and regulatory agency must have adequate information to insure that pollution will not occur and the facility will operate with minimal environmental degradation. EPA has issued a preliminary draft of a technical bulletin, "Evaluation of Land Application Systems."[1] This bulletin is to set forth procedures to assist EPA personnel in evaluating land treatment systems that are proposed in facility planning under Section 201 and to provide information and assistance that may be of value to other federal, state and local agencies, the wastewater industry, consultants, and designers.

EPA does not consider this to be a comprehensive design manual, nor does it appear that they intend to prepare a design manual, apparently because of the numerous variables present at each locality. Some of the variables that must be considered are:

1. composition and flow of wastewater
2. pretreatment needed to protect public health and to prevent nuisances
3. storage needed for the period that irrigation cannot be practiced
4. depth to groundwater
5. topography
6. climatic conditions
7. types of soils present and how they will react to long-term application of the constituents in the wastewater
8. type of grass or crops to be used
9. means of wastewater application and the rate of application
10. amount of land required and the number of parcels involved
11. treatment provided with various application rates, types of grasses and crops, and soils present
12. potential for surface water pollution through seepage, runoff, and erosion
13. potential for groundwater pollution and elevation of the water table
14. remoteness of the area and potential for encroachment upon this area
15. monitoring required to determine the effects on ground and surface waters
16. capital, operation, and maintenance costs
17. total environmental effects.

An evaluation checklist for treatment alternatives employing land application of wastewater taken from the EPA technical bulletin[1] is printed in its entirety in Appendix A because it shows the many items needed to appraise land application.

It is obvious that obtaining information to satisfy the items in the EPA checklist will be costly and could very well deter projects, particularly those of smaller size. However, the EPA technical bulletin does state:

> The scope of the Evaluation Checklist is aimed at moderate-to-large sized land-application systems. The extent to which planning and design of small systems (say 0.5 mgd or less) should adhere to all points in the checklist is left to the discretion of the evaluator.

In Montana, the most applicable situations for land disposal appear to be related to the small community using a stabilization pond in which the present effluent needs to be upgraded to meet secondary treatment standards. Many of these have adjacent land that may be suitable for land application.

It should be recognized by regulatory agencies that for a large project, the preliminary engineering and land acquisition will take considerable time, and regulations requiring treatment facilities to be constructed in a short period of time will probably rule out land treatment alternatives. This is true particularly where there is a need to acquire several parcels of land.

Groundwater

Soils present for treatment and groundwater conditions are the number one considerations in preliminary evaluations, and minimal investigations on these items will probably rule out many sites. When the total evaluation is completed, groundwater will undoubtedly be the item leaving the largest question mark. Montana, like most other states, does not have specific groundwater standards other than prohibition of pollution or contamination of existing domestic well water, stated in general terms.

An EPA document "Alternative Waste Treatment Techniques for Best Practicable Waste Treatment"[2] contains the following information:

> Once one alternative is selected, it must comply with certain additional requirements, described in this document. For example, any land application or land utilization techniques must, in order to qualify for Federal funding, comply with criteria designed to protect groundwaters. These criteria are intended to ensure that the nation's groundwater . . . resources remain suitable for drinking water purposes. The groundwaters in the zone of saturation in any aquifer resulting from land or subsurface disposal must meet the chemical and pesticide levels in the EPA public drinking water criteria "Manual for Evaluating Public Drinking Water." However, if the groundwater presently exceeds the specified quality, case by case exceptions may be allowed provided no further degradation ensues. If the land application technique results in a point source discharge to navigable waters—for example, one which utilizes an underdrain system—that discharge must comply with applicable effluent limitations for discharges from publicly owned treatment works.

If this becomes the final policy of EPA, land disposal could be unduly restricted. The critical point of measurements for monitoring a soil treatment system would be at the interface of percolating waste and groundwater with no mixing zone. A mixing zone for point source discharges into surface waters is provided by Montana and other states. Unless a mixing zone is allowed for groundwater discharges, many land treatment systems will be eliminated from consideration. Also, land disposal areas where the upper groundwater is already unfit for human consumption through natural conditions and where there is suitable water in a lower stratum that is amply protected from the upper water should not be ruled out because of a small increase in mineral content. Groundwater standards are needed to assure that good treatment is provided by the soils, but they should be developed realistically.

CONCLUSION

Public Law 92-500 requires the discharger to meet certain effluent limits and for a municipal discharger to consider various treatment alternatives (including land application techniques) in the preliminary planning. The EPA checklist for land treatment (Appendix A) presented should prove of assistance to the designer and the evaluator.

Groundwater quality could very well be the condition most difficult to meet and to assess fully in the preliminary work for land disposal. There may be groundwater standards developed by both the EPA and the states that could severely restrict the use of land disposal systems. In order to give ample protection of the ground and surface waters, land application systems must be designed as a treatment device and not merely as a means of disposal.

In the past, the civil-sanitary engineer has done essentially all the preliminary design work for a wastewater treatment project. It becomes apparent that other professions, including the soils scientist, the geologist, and the hydrogeologist, must also be involved in obtaining information for evaluation of a land treatment system.

REFERENCES

1. *Evaluation of Land Application Systems*, EPA-430/9-74-015, Office of Water Program Operations, Environmental Protection Agency (September 1974).
2. *Alternative Waste Treatment Techniques for Best Practicable Waste Treatment* (Draft), Office of Water Program Operations, Environmental Protection Agency (March 1974).

APPENDIX A

EVALUATION CHECKLIST FOR TREATMENT ALTERNATIVES EMPLOYING LAND APPLICATION OF WASTEWATER

This checklist is reproduced by permission from a preliminary draft of Technical Bulletin EPA 520/9-75-001. The document has not yet been published and is not to be construed to represent agency policy.

The purpose of this checklist is to provide reviewers with the pertinent factors to be considered in the planning, design, and operation of systems employing land application of municipal effluents. The format of the checklist has been selected to enable the reviewer to enter a check mark or comment to the right of each item. Items are arranged so that the more important ones appear first. Those items for which a dashed checkline appears are desirable but not essential considerations. The notation and headings used are generally the same as those used in the background information text.

Part I. Facilities Plan

A. Project Objectives

Objectives and goals relevant to water quality, protection of groundwater aquifer, the need for augmenting existing water resources, and any other desired effects should be considered initially.

B. Evaluation of Wastewater Characteristics

1. Flowrates
   Present, projected, and peak flow ________
2. Existing treatment
   a. Description ________
   b. Adequacy for intended project ________
3. Existing effluent disposal facilities
   a. Description ________
   b. Consideration of water rights ________
4. Composition of effluent to be applied
   a. Total dissolved solids ________
   b. Suspended solids ________
   c. Organic matter (BOD, COD, TOC) ________
   d. Nitrogen forms (all) ________
   e. Phosphorus ________
   f. Inorganic ions
      (1) Heavy metals and trace elements ________
      (2) Exchangeable cations (SAR) - - - - - -
      (3) Boron - - - - - -

I–B.4. (continued)

g. Bacteriological quality __________
h. Projected changes in characteristics __________
i. Are industrial wastewater components considered? __________
j. BPT constituents __________

C. Evaluation of Potential Sites

All potential sites should be considered on the basis of the criteria listed in this section, and should be reevaluated in the light of design considerations and environmental assessment.

1. General description

a. Location
(1) Distance from collection area or treatment plant __________
(2) Elevation relative to collection area __________

b. Compatibility with overall land-use plan
(1) Current use __________
(2) Proposed future use __________
(3) Zoning and adjacent land use __________
(4) Proximity to current and planned developed areas __________
(5) Is there room for future expansion? __________

c. Proximity to surface water __________

d. Number and size of available land parcels __________

2. Description of environmental characteristics

a. Climate
(1) Precipitation analysis and seasonal distribution __________
(2) Storm intensities __________
(3) Temperature, with seasonal variations __________
(4) Evapotranspiration __________
(5) Wind velocities and direction __________

b. Topography
(1) Ground slope __________
(2) Description of adjacent land __________
(3) Erosion potential __________
(4) Flood potential __________
(5) Extent of clearing and field preparation necessary __________

c. Soil characteristics
(1) Type and description __________
(2) Infiltration and percolation potential __________
(3) Soil profile __________
(4) Evaluation by soil specialists __________

I–C.2. (continued)

d. Geologic formations
- (1) Type and description ________
- (2) Evaluation by geologist ________
- (3) Depth of formations ________
- (4) Earthquake potential ________

e. Groundwater
- (1) Depth to groundwater ________
- (2) Groundwater flow ________
- (3) Depth and extent of any perched water ________
- (4) Quality compared to requirements ________
- (5) Current and planned use ________
- (6) Location of existing wells
  - (a) On-site ________
  - (b) Adjacent to site ________

f. Receiving water (other than groundwater)
- (1) Type of body ________
- (2) Current use ________
- (3) Existing quality ________
- (4) Is it water-quality limited? ________
- (5) Is it effluent limited? ________
- (6) Water rights ________

3. Methods of land acquisition or control

- a. Purchase ________
- b. Lease ________
- c. Purchase and lease back to farmer ________
- d. Contract with users ________
- e. Other ________

D. Consideration of Land-Application Alternatives

Based on the project objectives and characteristics of the selected potential sites, appropriate methods of land application should be considered.

1. Irrigation

a. Purpose
- (1) Optimization of crop yields ________
- (2) Maximization of effluent application ________
- (3) Landscape irrigation ________

b. Application techniques
- (1) Spraying ________
- (2) Ridge and furrow ________
- (3) Flooding ________

2. Infiltration-percolation

a. Purpose
- (1) Groundwater recharge ________
- (2) Pumped withdrawal or underdrains ________
- (3) Interception by surface water ________

I–D.2. (continued)

- b. Application techniques
  - (1) Spreading ________
  - (2) Spraying ________
3. Overland flow (spray-runoff)
   - a. Purpose
     - (1) Discharge to surface waters ________
     - (2) Reuse of collected runoff ________
   - b. Application techniques
     - (1) Spraying ________
     - (2) Flooding ________
4. Combinations of treatment techniques
   - a. Combinations of land-application techniques at the same or different sites ------
   - b. Combinations of land-application with in-plant treatment and receiving water discharge ------
5. Compatibility with site characteristics ________

E. Design Considerations

1. Loading rates
   - a. Liquid loading/water balance
     - (1) Design precipitation ________
     - (2) Effluent application ________
     - (3) Evapotranspiration ________
     - (4) Percolation ________
     - (5)* Runoff (for overland flow systems) ________
   - b. Nitrogen mass balance
     - (1) Total annual load ________
     - (2) Total annual crop uptake ________
     - (3) Denitrification and volatilization ________
     - (4) Addition to groundwater or surface water ________
   - c. Phosphorus mass balance ________
   - d. Organic loading rate (BOD)
     - (1) Daily loading ________
     - (2) Resting-drying period for oxidation ________
   - e. Loadings of other constituents ________
2. Land requirements
   - a. Field area requirement ________
   - b. Buffer zone allowance ________
   - c. Land for storage ________
   - d. Land for buildings, roads and ditches ________
   - e. Land for future expansion or emergencies ________

---

*(6) Need for sub-drains [Eds.]

I–E. (continued)

3. Crop selection
   a. Relationship to critical loading parameter ______
   b. Public health regulations ______
   c. Ease of cultivation and harvesting ______
   d. Length of growing season ______
   e. Landscape requirements ______
   f. Forestland ______
4. Storage requirements
   a. Related to length of operating season and climate ______
   b. For system backup ______
   c. For flow equalization ______
   d. Secondary uses of stored wastewater ______
5. Preapplication treatment requirements
   a. Public health considerations ______
   b. Relationship to loading rate ______
   c. Relationship to effectiveness of physical equipment ______
6. Management considerations
   a. System control and maintenance ______
   b. Manpower requirements ______
   c. Monitoring requirements ______
   d. Emergency procedures ______
7. Cost-effectiveness analysis
   a. Capital cost considerations
      (1) Construction or other cost index ______
      (2) Service life of equipment ______
      (3) Land cost ______
   b. Fixed annual costs
      (1) Labor ______
      (2) Maintenance ______
      (3) Monitoring ______
   c. Flow-related annual costs
      (1) Power ______
      (2) Crop sale or disposal ______
   d. Nonmonetary factors ______
8. Flexibility of alternative
   a. With regard to changes in treatment requirements ______
   b. With regard to changes in wastewater characteristics ______
   c. For ease of expansion ______
   d. With regard to changing land utilization ______
   e. With regard to technological advances ______
9. Reliability
   a. To meet or exceed discharge requirements ______
   b. Failure rate due to operational breakdown ______
   c. Vulnerability to natural disasters ______
   d. Adequate supply of required resources ______
   e. Factors-of-safety ______

I–E. (continued)

10. Best practicable waste treatment technology (BPT)
    a. Requirements for groundwater quality ________
    b. Requirements for treatment and discharge ________

F. Environmental Assessment

The impact of the project on the environment, including public health, social, and economic aspects must be assessed for each land-application alternative.

1. Environmental impact
    a. On soil and vegetation ________
    b. On groundwater
        (1) Quality ________
        (2) Levels and flow direction ________
    c. On surface water
        (1) Quality ________
        (2) Influence on flow ________
    d. On animal and insect life ________
    e. On air quality ________
    f. On local climate ________
2. Public health effects
    a. Groundwater quality ________
    b. Insects and rodents ________
    c. Runoff from site ________
    d. Aerosols ________
    e. Contamination of crops ________
3. Social impact
    a. Relocation of residents ________
    b. Effects on greenbelts and open space ________
    c. Effect on recreational activities ________
    d. Effect on community growth ________
4. Economic impact
    a. On overall local economy ________
    b. Tax considerations (land) ________
    c. Conservation of resources and energy ________

G. Implementation Program

The ability to implement the project must be assessed in light of the overall impact, the effectiveness of the tentative design, and with regard to public opinion.

1. Public information program
    a. Approaches to public presentation
        (1) Local officials ________
        (2) Public hearings ________

I–G.1. (continued)

- (3) Mass media ________
- (4) Local residents and land owners ________
- (5) Communication with special-interest groups ________

b. Public opinion

- (1) Engineer's response ________
- (2) Review of problem areas ________

2. Legal considerations ________
3. Reevaluation of ability to implement project ________
4. Implementation schedule
   - a. Construction schedule ________
   - b. Long-range management plan ________

2

USE OF THE SOIL-VEGETATION BIOSYSTEM FOR WASTEWATER RECYCLING

William E. Sopper

Professor of Forest Hydrology
School of Forest Resources
The Pennsylvania State University
University Park, Pa. 16802

INTRODUCTION

Many water pollution problems have been created by the disposal of treated municipal wastewaters into streams, lakes, and oceans. There are currently about 16,000 sewage treatment plants in the United States discharging over 26 billion gallons of effluent daily. These wastewaters, although acceptable by public health requirements, are usually enriched with appreciable quantities of dissolved minerals and synthetic detergent residues. As a result, the concentrated discharge of large volumes of these effluents into a balanced aquatic environment, often causes ecological chaos and disrupts the natural recycling process. The visible evidence is usually eutrophication and stimulation of the growth of aquatic plants. In some instances, even fish kills may result from reduced dissolved oxygen levels in the water resulting from respiration by these plants and decomposition of organic matter.

Increasing volumes of municipal wastewater are usually correlated with increasing demands on the local water supply which in times of drought can cause serious water shortages. Hence, it is somewhat paradoxical that communities while experiencing water shortages will at the same time discharge millions of gallons of wastewater into local streams for rapid removal from the area.

An obvious alternate method to disposal of sewage effluent in surface waters is to dispose of such effluents on the land so as to utilize the entire biosystem, soil and vegetation, as a "living filter" to renovate this effluent for groundwater recharge. Under controlled application rates to maintain aerobic conditions within the soil, the mineral nutrients and detergent residual might be removed and degraded by: (1) microorganisms in the surface soil horizons, (2) chemical precipitation, (3) ion exchange, (4) biological transformation, and (5) biological absorption through the root systems of the vegetative cover. The utilization of the higher plants as an integral part of the system to complement the microbiological and physiochemical systems in the soil is an essential component of the living filter concept and provides maximum renovation capacity and durability to the system.

Water pollution and water supply problems in State College, Pennsylvania, are typical of those in many communities. Discharge of increasing volumes of wastewater from its sewage treatment plant resulted in pollution of Spring Creek, the area's main water course. In addition, the area's water supply, which comes solely from the groundwater reservoir, was gradually diminishing from a prolonged drought that started in 1960.

The necessity of finding solutions to both problems led to a full-scale investigation into the feasibility of land disposal for large volumes of wastewater and the living filter concept. In 1962, an interdisciplinary team consisting of agricultural and civil engineers, agronomists, foresters, geologists, microbiologists, biochemists, and zoologists was assembled to attack the problem.

A pumping house, which was capable of delivering 500,000 gallons of chlorinated effluent per day to two disposal sites, 3.2 and 6.4 km away, on the University farms, was constructed at the treatment plant. Sprinkler irrigation systems were established at the two sites to irrigate approximately 28 hectares of forest and crop land. Details of the system have been previously described by Parizek, *et al.*[1]

Treated municipal sewage effluent has been spray irrigated on cropland and in forested areas for a 12-year period (1963-1974) at the Penn State Project. The results of this research will be used to illustrate some of the benefits and evils of using a soil-vegetation biosystem for wastewater recycling.

Forested areas irrigated consisted of a mixed hardwood forest, a red pine plantation (*Pinus resinosa*), and a sparse white spruce (*Picea glauca*) plantation established on an abandoned old field. Types of crops irrigated were wheat, oats, corn, alfalfa, red clover, and reed canarygrass. Detailed descriptions of these areas have been previously reported by Sopper.[2]
The two soil types present on the site are the Hublersburg, with a surface

texture ranging from silt loam to silty clay loam on slopes ranging from 3 to 12%, and the Morrison sandy loam with slopes ranging from 2 to 20%.

Sewage effluent was applied in varying amounts, from 2.5 cm per week to 10 cm per week, and over varying lengths of time, from 16 weeks during the growing season to the entire 52 weeks. Rates of application varied from 6.2 to 16 mm per hour.

CHEMICAL COMPOSITION OF MUNICIPAL SEWAGE TREATMENT PLANT EFFLUENT

The chemical composition of municipal effluent is illustrated in Table 2.1 based upon samples collected from the University treatment plant. All wastewater is given both primary and secondary treatment, with secondary treatment given by either standard or high-rate trickling filters followed by a modified activated sludge process and final settling. The total amount of each constituent applied per hectare per year at the 5 cm per week application rate is also given in Table 2.1.

The fertilizer value of these wastewaters is readily evident in that the 5 cm per week application provides commercial fertilizer constituents equivalent to approximately 233 kilograms per hectare of nitrogen, 224 kilograms of phosphate (P_2O_5), and 254 kilograms of potash (K_2O). This is equivalent to applying about 2200 kilograms of a 10-10-10 fertilizer per hectare annually, which is more than twice as great as a "normal" application of commercial fertilizer.

RENOVATION

Crop Areas

Nitrogen and phosphorus are the two key eutrophic elements in municipal wastewater and therefore discussion on renovation will be limited to these two elements. The overall efficiency of the biological system to accept and renovate wastewater can be evaluated by observations on the quality of percolating water within the surface soil mantle. Suction lysimeters installed in all areas were used to obtain samples of percolating water at the 120 cm soil depth. Percolate samples were collected, whenever possible, after each weekly application of wastewater and analyzed for the same constituents as the wastewater.

Mean annual concentrations of phosphorus and nitrate-nitrogen for the agronomic areas during the period 1965 to 1973 are given in Tables 2.2 and 2.3. Mean annual phosphorus concentrations in the sewage effluent

Table 2.1. Typical Chemical Composition of Treated Municipal Sewage Effluent

Constituent	Average Concentration mg/l	Total amount applied[a] kg/ha
pH	8.1	–
MBAS[b]	0.37	6
Nitrate-N	8.6	143
Organic-N	2.4	40
NH_4-N	0.9	14
P	2.651	44
Ca	25.2	420
Cl	41.3	792
Mg	12.9	215
Na	28.1	469
Fe	0.4	9
	μg/l	kg/ha
B	169	3.26
Mn	61	1.15
Cu	109	1.96
Zn	211	4.15
Cr	23	0.41
Pb	104	2.12
Cd	9	0.19
Co	62	1.24
Ni	93	1.82

[a]Total amount applied on areas that received 5 cm of effluent per week

[b]Methylene blue active substance (detergent residue).

have varied from 2.5 to 10 mg/l. Phosphorus concentrations in the percolating water at the 120 cm soil depth have been consistently decreased by more than 98% since the initiation of the project in 1963. Even when application rates were increased from 5 to 7.5 cm per week in 1972-73 the degree of renovation was essentially unchanged. The differences between the mean annual concentrations of phosphorus on the control and irrigated areas are quite small and insignificant considering that more than 13 meters of sewage effluent have been applied over the 11 years. It should also be noted that starting in year 1971 the reed canarygrass area has been irrigated with a mixture of sewage effluent and liquid digested

Table 2.2. Phosphorus in Percolate from Croplands.
Mean annual concentration of phosphorus in suction lysimeter samples collected at the 120-cm soil depth in the corn rotation plots and the reed canarygrass area, mg/l

Year	Corn cm per week		Reed Canarygrass cm per week	
	0	5	0	5
1965	0.032	0.022	–	–
1966	0.045	0.036	–	0.055
1967	0.039	0.054	–	0.053
1968	0.041	0.060	–	0.052
1969	0.066	0.070	–	0.035
1970	0.034	0.075	–	0.038
1971	0.048	0.061	–	0.061
1972[a]	0.022	0.035	0.067	0.054
1973[a]	0.013	0.020	0.036	0.052

[a]Application rate increased to 7.5 cm per week.

Table 2.3. Nitrogen in Percolate from Croplands.
Mean annual concentration of nitrate-nitrogen in suction lysimeter samples collected at the 120-cm soil depth in the corn rotation plots and the reed canarygrass area, mg/l

Year	Corn cm per week		Reed Canarygrass cm per week	
	0	5	0	5
1965	5.2	9.7	–	–
1966	4.7	7.0	–	3.7
1967	3.4	7.1	–	3.3
1968	4.5	9.5	–	3.1
1969	9.4	13.5	–	2.5
1970	10.3	10.9	–	2.4
1971	6.2	9.6	–	3.3
1972[a]	9.4	10.6	1.7	7.7
1973[a]	7.4	7.4	1.5	9.2

[a]Application rate increased to 7.5 cm per week.

sludge (approximately 1 part sludge slurry to 11 parts effluent). This effluent-sludge mixture had a total phosphorus concentration of 20 to 25 mg/l and an orthophosphate level of 10 to 15 mg/l. These data indicate that phosphorus is not leaching out of the soil profile into the groundwater at significantly higher concentrations from the wastewater-treated areas than from the control areas and that there appears to be no decreasing trend evident in terms of satisfactory phosphorus removal.

Nitrate-nitrogen renovation was not as efficient on the agronomic areas. Nitrate-nitrogen concentrations were decreased below the 10 mg/l level recommended by the Public Health Service for drinking water on the corn rotation area that received the 2.5 cm per week treatment but not at the 5 cm per week level. At the higher application rate (5 cm) the mean annual concentration remained below the PHS limit only when grass-legume hays occupied 28 to 68% of the site during the period 1965 to 1968. Starting with 1969 the entire site has been planted with corn. Increases in the application rate from 5 to 7.5 cm during 1972 and 1973 did not significantly affect the degree of renovation on the corn area.

On the other hand, the 5 cm per week reed canarygrass area has been exceptionally efficient in nitrogen removal (with the canarygrass harvested three to four times annually) and has consistently maintained the mean annual concentration of nitrate-nitrogen in percolating water below the 10 mg/l limit. This area is irrigated year-around and therefore received twice as much nitrogen as the corn area. Even when application rates were increased to 7.5 cm per week during 1972 and 1973, mean annual concentrations were increased somewhat but still remained within acceptable levels.

Forest Areas

The forest areas were highly efficient in removing phosphorus (Table 2.4). Mean annual concentrations are slightly higher than those of the agronomic areas because there is no harvested crop and the phosphorus is continually recycled. Year-around irrigation of the forest area on the Morrison soil at the 5 cm per week level has resulted in a substantial increase in phosphorus concentration in the soil percolate. As shown in Table 2.4, there has been a steady increase in comparison to the control area since the third year (1968) of operation. On the other hand, phosphorus concentrations in the irrigated forest areas on the Hublersburg soil, which received lower application rates and shorter periods of irrigation (growing season, 30 to 33 weeks), were similar to those in the control areas. Increases in the application rates from 2.5 to 3.8 cm and from 5 to 7.5 cm during the growing season irrigation period did not appear to reduce renovation efficiency.

Table 2.4. Phosphorus in Percolate from Forests.
Mean annual concentration of phosphorus in suction lysimeter samples collected at the 120-cm soil depth in the forest areas, mg/l

Year	Red Pine, Hublersburg Soil, cm per week		Hardwood Hublersburg Soil, cm per week		Old Field Hublersburg Soil, cm per week		Hardwood Morrison Soil, cm per week	
	0	2.5	0	2.5	0	5	0	5
1965	0.040	0.300	0.050	0.250	–	0.460	–	–
1966	0.043	0.134	0.037	0.043	0.030	0.140	0.059	0.042
1967	0.053	0.092	0.044	0.077	0.039	0.068	0.068	0.063
1968	0.075	0.089	0.106	0.222	0.040	0.053	0.071	0.116
1969	0.010	0.064	0.072	0.047	0.051	0.098	0.059	0.137
1970	0.065	0.076	0.033	0.143	0.042	0.114	0.051	0.209
1971	0.994	0.107	0.015	0.146	0.039	0.420	0.037	0.378
1972	0.113	0.105[a]	0.078	0.037[a]	0.074	0.200[b]	0.033	0.335
1973	0.028	0.035[a]	0.022	0.051[a]	0.042	0.096[b]	0.020	0.392

[a]Application rate increased to 3.8 cm per week.

[b]Application rate increased to 7.5 cm per week.

The efficiency of forest areas to reduce nitrogen concentrations has been variable. It appears that forest areas can accept a 2.5 cm per week application of wastewater without having the mean annual concentration of nitrate-nitrogen exceed the PHS limit (Table 2.5). However, year-around irrigation of forests at 5 cm per week consistently resulted in no renovation. It also appears that forest ecosystems may be quite sensitive to wastewater application rates and may have a low threshold in terms of collapse in the renovation system. A 50% increase in the application rates (2.5 to 3.8 cm and 5 to 7.5 cm) after nine years of successful operation and renovation resulted in a complete breakthrough in terms of concentrations of nitrate-nitrogen in percolating water. Mean annual concentrations nearly tripled the first year (1972) and all exceeded 10 mg/l.

The Old Field area has been somewhat exceptional in comparison to the other forest areas in terms of nitrogen renovation. This area receives the highest application rate on the Hublersburg soil and yet has consistently maintained nitrate-nitrogen levels below 10 mg/l through the nine year period (1963-1971) without the harvest of any crop. In 1963 at the start of the project the area was primarily an open weed field with a few scattered spruce saplings (1 to 2 meters in height). Although the trees are now more than 7 meters in height, the spruce stand is still sparse with fairly large open areas. It appears that the annual and perennial weeds that occupy these open areas during the growing season (irrigation period) provide a temporary storage for nitrogen, thereby reducing nitrate-nitrogen leaching losses. In the fall, vegetative growth ceases and nitrates are again available for leaching. However, since irrigation has ceased by this time, the concentrations of nitrate-nitrogen in percolate water remains at a low level.

This same phenomenon was observed in the red pine plantation, which was irrigated at the 5 cm per week level. Mean annual concentrations of nitrate-nitrogen steadily increased from 1963 to 1969 (Table 2.6). In November, 1968 a snow storm resulted in complete blowdown of the plantation. In 1969 the area was clearcut and all trees were removed. Immediately a dense cover of herbaceous vegetation developed, similar to that on the irrigated Old Field area. As a result the mean annual concentration of nitrate-nitrogen decreased from 24.2 mg/l in 1969 to 8.3 mg/l in 1970 and to 2.9 mg/l in 1971. Subsequent increases in 1972 and 1973 were due to the increased application rate.

EFFECTS ON CHEMICAL PROPERTIES OF THE SOIL

Soil samples were taken on all areas to a depth of 150 cm in the fall after cessation of irrigation in 1963, 1967, and 1971. Soil samples were

Table 2.5. Nitrogen in Percolate from Forests.
Mean annual concentration of nitrate-nitrogen in suction lysimeter samples collected at the 120-cm soil depth in the forest areas, mg/l

	Red Pine, Hublersburg Soil cm per week		Hardwood Hublersburg Soil cm per week		Old Field Hublersburg Soil cm per week		Hardwood Morrison Soil cm per week	
Year	0	2.5	0	2.5	0	5	0	5
1965	0.9	2.2	–	0.0	0.3	8.0	–	–
1966	0.1	2.1	0.1	0.2	0.1	5.0	0.1	10.6
1967	0.9	1.7	0.3	1.4	0.3	6.1	1.4	19.2
1968	0.9	2.7	0.1	8.0	0.2	3.7	0.1	25.9
1969	0.2	4.2	0.1	7.2	0.2	2.3	0.3	23.7
1970	<1	5.3	<1	5.0	<1	3.5	1.0	42.8
1971	2.6	8.3	0.5	5.8	0.5	3.8	0.8	17.6
1972	6.0	21.8[a]	14.7	23.9[a]	3.2	11.8[b]	4.7	22.9
1973	0.5	13.7[a]	3.7	13.5[a]	0.5	13.5[b]	1.3	17.3

[a]Application rate increased to 3.8 cm per week.

[b]Application rate increased to 7.5 cm per week.

Table 2.6. Nitrogen in Percolate from Red Pine.
Mean annual concentration of nitrate-nitrogen in suction lysimeter samples collected at the 120-cm soil depth in the red pine plantation irrigated at 5 cm per week, mg/l

Year	Red Pine Hublersburg Soil cm per week	
	0	5
1965	0.9	3.9
1966	0.1	9.3
1967	0.9	13.8
1968	0.9	19.9[a]
1969	0.2	24.2[b]
1970	<1	8.3
1971	2.6	2.9
1972	6.0	14.5[c]
1973	0.5	9.0[c]

[a]Blowdown in November, 1968.

[b]All pine trees cut and removed.

[c]Application rate increased to 7.5 cm per week.

analyzed for the same constituents as was the effluent to determine if any significant concentrations of nutrients were accumulating in the irrigated areas.

The exchangeable cations (K, Ca, Mg, Na, and Mn) and boron were extracted from the soil with ammonium acetate[3] and analyzed with an arc spectrometer.[4] Exchangeable hydrogen was determined using a barium chloride buffering technique.[3] Organic matter (OM) was determined using a potassium dichromate-sulfuric acid oxidation method.[5] Soil pH was measured with a glass electrode using a 1:1 soil to water mixture. Total nitrogen (N) was analyzed using modified Kjeldahl method to include nitrates,[3] and the phosphorus concentration was determined by using the Bray extraction procedure.[3]

The results of the soil analyses for 1963 and 1971 for the two areas on each soil type that received the highest applications of effluent are given in Tables 2.7 and 2.8. These two areas were the Old Field 5-cm area on the Hublersburg soil and the hardwood 5-cm area on the Morrison soil.

Results of the soil analyses of the 1971 samples indicated that the effects of effluent irrigation on exchangeable potassium, organic matter, pH and total nitrogen were small and inconsistent. However, there were significant

Table 2.7. Concentrations in Old Field
Mean constituent concentration for the Old Field 5-cm area on the Hublersburg soil type

Year and Area	Depth cm	me/100 g K	Ca	Mg	Na	H	ppm Mn	B	P	pH	%N	%OM
1963 Control	30	0.40	1.43	0.27	0.10	11.84	71.60	0.53	8.70	4.79	0.088	2.147
	60	0.47	2.00	0.83	0.17	8.67	20.46	0.60	0.76	5.23	0.016	0.370
	90	0.47	1.13	1.27	0.20	9.33	23.46	0.53	0.38	4.91	0.012	0.167
	120	0.43	0.93	1.57	0.20	–	18.60	0.53	0.38	–	0.010	–
	150	0.47	0.97	1.53	0.13	–	18.13	0.53	0.38	–	0.010	–
1963 Treatment	30	0.63	1.73	0.50	0.23	16.06	44.60	1.00	16.05	4.96	0.072	2.407
	60	0.60	1.87	0.90	0.20	18.74	20.80	0.80	1.48	4.68	0.022	0.390
	90	0.63	1.03	1.00	0.10	11.48	26.00	0.86	0.71	4.45	0.019	0.183
	120	0.53	0.77	0.97	0.10	–	62.00	0.80	0.00	–	0.014	–
	150	0.47	0.57	0.77	0.10	–	17.93	0.80	0.00	–	0.016	–
1971 Control	30	0.10	2.00	0.23	0.23	14.91	15.86	0.06	10.95	5.06	0.082	2.683
	60	0.03	2.16	0.80	0.23	11.56	12.13	0.06	0.91	4.89	0.016	0.313
	90	0.00	1.80	1.43	0.23	12.05	10.60	0.00	1.00	4.84	0.009	0.167
	120	0.03	1.53	1.63	0.20	–	9.46	0.06	0.90	–	0.005	–
	150	0.00	1.16	1.50	0.20	–	8.66	0.00	1.78	–	0.005	–
1971 Treatment	30	0.43	3.33	1.73	0.33	15.52	14.53	0.60	50.80	5.42	0.095	2.300
	60	0.20	2.00	1.33	0.30	16.36	17.20	0.20	3.48	4.99	0.026	0.343
	90	0.06	1.40	1.23	0.33	16.68	15.80	0.06	1.90	4.82	0.017	0.190
	120	0.16	1.00	1.36	0.30	–	15.66	0.06	1.66	–	0.020	–
	150	0.13	1.00	1.43	0.33	–	16.06	0.13	1.76	–	0.028	–

Table 2.8. Concentrations in Hardwood Forest
Mean constituent concentrations for the hardwood 5-cm area on the Morrison soil type

	Depth	me/100 g					ppm					
Year and Area	cm	K	Ca	Mg	Na	H	Mn	B	P	pH	%N	%OM
1967 Control	30	0.16	0.90	0.10	0.16	13.43	55.93	0.26	16.50	4.55	–	2.210
	60	0.16	0.70	0.33	0.16	12.07	16.46	0.13	2.13	4.67	–	0.547
	90	0.16	0.46	0.33	0.20	10.38	10.46	0.13	1.23	4.76	–	0.223
	120	0.26	0.43	0.53	0.20	–	4.80	0.33	0.83	–	–	–
	150	0.06	0.36	0.56	0.20	–	10.13	0.06	0.88	–	–	–
1967 Treatment	30	0.13	1.36	0.40	0.23	12.94	19.93	0.26	75.55	5.50	–	1.240
	60	0.06	0.50	0.30	0.23	10.76	11.86	0.00	4.90	4.99	–	0.560
	90	0.13	0.46	0.36	0.23	13.37	19.20	0.13	2.18	4.94	–	0.350
	120	0.16	0.60	0.36	0.20	–	16.26	0.13	1.92	–	–	–
	150	0.26	1.46	1.10	0.26	–	15.33	0.20	1.50	–	–	–
1971 Control	30	0.00	0.53	0.10	0.20	13.72	32.73	0.06	6.25	4.96	0.109	2.867
	60	0.16	1.13	0.80	0.20	13.36	11.00	0.20	3.16	4.97	0.036	0.300
	90	0.26	0.76	1.03	0.20	11.54	10.86	0.26	0.66	4.99	0.028	0.230
	120	0.20	0.53	0.83	0.20	–	9.86	0.20	1.06	–	0.023	–
	150	0.00	0.50	0.96	0.16	–	7.26	0.00	0.68	–	0.016	–
1971 Treatment	30	0.06	1.86	0.53	0.23	7.35	4.13	0.13	143.75	6.15	0.084	1.483
	60	0.03	0.96	0.36	0.20	5.20	4.93	0.20	31.60	5.51	0.052	0.413
	90	0.10	1.30	0.73	0.20	9.34	5.73	0.13	9.95	5.06	0.037	0.243
	120	0.03	1.10	0.83	0.23	–	6.40	0.06	1.98	–	0.032	–
	150	0.13	0.73	0.80	0.23	–	4.20	0.20	0.98	–	0.032	–

changes in the concentrations of calcium, magnesium, sodium, manganese, boron, and phosphorus. Calcium, magnesium, and boron concentrations increased significantly at the 30-cm depth in both the Hublersburg and Morrison soils and in all the vegetative cover types. Sodium concentrations increased significantly at all depths in the Hublersburg and Morrison soils and in all the vegetative cover types. Manganese concentrations decreased significantly in the upper 90 cm of the Hublersburg soil on the hardwood 2.5 treated plot and the upper 90 cm of the Morrison soil on the new gameland hardwood 5-cm treated plot. Phosphorus concentrations increased significantly in the Hublersburg soil in the 30-cm depth of the hardwood 2.5 cm plot, the upper 60 cm of the Old Field 5-cm and red pine 2.5 cm plots. Phosphorus also increased significantly in the upper 150 cm of the Morrison soil on the new gameland hardwood 5-cm plot.

Of the 11 constituents analyzed, only potassium, sodium, manganese, exchangeable hydrogen, boron, and phosphorus had significant changes over time. Potassium concentrations decreased in all five depths of the treated and control plots in the Hublersburg soil on the Old Field 5-cm plots. Sodium concentrations increased significantly in all five depths of the Hublersburg soil in the Old Field 5-cm treated plot, the 60 and 90 cm depths of the hardwood 2.5-cm treated plot, and the lower 90 cm of the red pine 2.5-cm treated plots. Manganese concentrations decreased significantly in all five depths on both soil types and in all vegetation cover types. Exchangeable hydrogen concentrations decreased significantly in the upper 90 cm of the Morrison soil on the new gameland hardwood 5-cm treated plot. Boron decreased significantly in all five depths of the Hublersburg soil on the Old Field 5-cm and red pine 2.5-cm treated plots. Phosphorus increased significantly in the 30 cm of the Hublersburg soil on the hardwood 2.5-cm and red pine 2.5-cm treated plots, and in the upper 90 cm of the Hublersburg soil on the Old Field 5-cm and the Morrison soil on the new gameland hardwood 5-cm treated plots.

CROP RESPONSES

Yields

During the initial years of the project a variety of crops were tested. Average annual crop yields during the period 1963 to 1970 have previously been reported by Sopper and Kardos.[6] Since 1968 the two primary crops grown have been silage corn and reed canarygrass. Our experience has shown that these two crops are the best suited to our site and are the most efficient in terms of nutrient utilization. During the past 12 years the crop areas irrigated with 5 cm of effluent weekly have received a total of approximately 20 meters of wastewater. During this period annual yield increases have ranged from 0 to 350% for corn grain, 5 to 130% for corn silage, 85 to 190% for red clover, and 79 to 140% for alfalfa.

Nutrient Composition

Under the "living filter" concept the vegetative cover is an integral part of the system and should complement the microbiological and physio-chemical activities occurring within the soil to renovate the effluent by removal and utilization of the nutrients applied. The crops harvested from the irrigated areas are usually higher in nitrogen and phosphorus than the control crops. However, the differences are not large (Tables 2.9 and 2.10). This is partially because the control area receives a normal application of commercial fertilizer each year, equal to about 900 kilograms of a 10-10-10 fertilizer per hectare annually.

Table 2.9. Nutrient Composition of Corn.
Average nutrient composition of corn silage Pa. 602-A receiving various levels of effluent, per cent

	Amount of Effluent Applied per Week		
Nutrient	0	2.5	5
Nitrogen	1.34	1.36	1.34
Phosphorus	0.21	0.28	0.35
Potassium	1.04	1.10	1.07
Calcium	0.48	0.32	0.23
Magnesium	0.16	0.21	0.19
Chloride	0.19	0.36	0.39
	μg/g	μg/g	μg/g
Sodium	12	167	201
Boron	7	7	7

Nutrients Removed by Crop Harvest

The contribution of the higher plants as renovators of the wastewater is readily evident when one considers the quantities of nutrients, expressed in kilograms per hectare, removed in crop harvest. Such data indicate that the vegetative cover can contribute substantially to the durability of a

Table 2.10. Nutrient Composition of Reed Canarygrass. Average nutrient composition of reed canarygrass irrigated with 5 cm of effluent weekly, per cent

Nutrient	Average Composition of Three Cuttings
Nitrogen	3.69
Phosphorus	0.50
Potassium	2.23
Calcium	0.40
Magnesium	0.36
Chloride	1.57
	μg/g
Sodium	309
Boron	8

"living filter" system particularly where a crop is harvested and utilized. At the 5-cm per-week level of effluent irrigation, the harvest of corn silage removes about 179 kilograms of nitrogen and 48 kilograms of phosphorus. Reed canarygrass, which is a perennial grass, is even more efficient in that it removes about 457 kilograms of nitrogen and 63 kilograms of phosphorus. The difference results primarily because the grass is already established and actively growing in early spring even before the corn is planted. The amounts of nutrients removed annually vary with the amount of wastewater applied, amount of rainfall, length of the growing season, and the number of cuttings of the reed canarygrass.

The efficiency of crops as renovating agents can be assessed by computing a "removal efficiency" expressed as the ratio of the weight of the nutrient removed in the harvested crop to the same nutrient applied in the wastewater. Average renovation efficiencies for the silage corn and the reed canarygrass crops are given in Table 2.11. At the 2.5-cm per week level of application of wastewater, the corn silage removes nutrients equivalent to about 334% of the total applied nitrogen, 230% of the applied phosphorus, and 280% of the applied potassium. At the 5-cm per week level, the corn silage removed more than 100% of the applied nitrogen, phosphorus, and potassium.

Similarly harvest of reed canarygrass removes on the average about 373 kilograms of nitrogen and about 50 kilograms of phosphorus. These removals are equivalent to renovation efficiencies of 122 and 63% respectively.

Table 2.11. Renovation Efficiency.
Average renovation efficiency of the silage corn and reed canarygrass crops, per cent

	Amount of Effluent Applied		
	Corn Silage Pa. 602-A		Reed Canarygrass
Nutrient	**2.5**	**5**	**5**
Nitrogen	334	145	122
Phosphorus	230	143	63
Potassium	280	130	117
Calcium	38	15	9
Magnesium	53	27	19
Chloride	26	14	20
Sodium	2	1	1
Boron	10	4	2

TREE GROWTH RESPONSES

Red Pine

Experimental plots were established in a red pine plantation in 1963. These plots have been irrigated with sewage effluent during the past 12 years at rates of 2.5 cm and 5 cm per week during the growing season (April to November). The plantation was established in 1939 with the trees planted at a spacing of 2.5 by 2.5 meters. In 1963 the average tree diameter at breast height was 22 cm and average height was 11 meters.

Diameter and height growth measurements were made annually. Average annual height growth on the 2.5-cm per week plot was 58 cm compared to 42 cm on the control plot. On the plot receiving 5 cm per week, height growth continually decreased up to 1968 when high winds following a wet snowfall completely felled every tree on the plot.

Increment cores were taken from sample trees in all areas to determine average annual diameter growth. Irrigation at the 2.5-cm per week level resulted in an average annual diameter growth of 4.3 mm in comparison to 1.5 mm on the control, an increase of 186%.

White Spruce

Two experimental plots were established in a sparse white spruce plantation on an abandoned Old-Field area. The trees in 1963 ranged from 0.9 to 2.5 meters in height. One plot has been irrigated with sewage effluent during the past 12 years at the rate of 5 cm per week, while the second plot has been maintained as a control.

Average height of the trees on the irrigated plot in 1974 was 7.5 meters and ranged from 5.5 to 9.5 meters. The average height of the trees on the control plot was 3.1 meters and ranged from 2.0 to 4.8 meters. Over the 12-year period average annual height growth was 60 cm on the irrigated areas and 25 cm on the control areas, representing a 140% increase as a result of sewage effluent irrigation.

Average diameter of trees on the irrigated plot was 9.5 cm in comparison to 3.6 cm on the control plot. Measurements taken from increment cores indicated that the average annual diameter growth on the irrigated trees was 10 mm and on the control trees 4.5 mm representing a 122% increase.

Mixed Hardwoods

A hardwood forest, consisting primarily of oak species, was irrigated with sewage effluent at 2.5 cm per week during the growing season and at 10 cm per week for the entire year (52 weeks).

Average annual diameter growth during the 1963 to 1974 period is given in Table 2.12. Application at 2.5 cm per week produced only a slight increase in diameter growth; however, the 5-cm per week level resulted in an 80% increase. These values pertain primarily to the oak species. Some of the other hardwood species present on the plots have responded to a greater extent. For example, increment core measurements made on red maple (*A. rubrum*) and sugar maple (*A. saccharum*) indicate that the average diameter growth during the past 12 years has been 13 mm on the trees irrigated with 2.5 cm of effluent per week in comparison to 2.6 mm on control trees, a 400% increase in average annual diameter growth.

Table 2.12. Growth of Trees.
Average annual diameter growth in hardwood forests irrigated with sewage effluent.

Weekly Irrigation Amount, cm	Average Diameter Growth	
	Control, mm	Irrigated, mm
2.5[a]	4.1	4.8
5[b]	3.3	6.0

[a]Irrigated with 2.5 cm of sewage effluent weekly during growing season from 1963 to 1974.

[b]Irrigated with 5 cm of sewage effluent weekly during the entire year from 1965 to 1974.

RENOVATION EFFICIENCY OF FORESTS

The nutrient element content of the foliage of the vegetation on the irrigated plots was consistently higher than that of the vegetation on the control plots. It is therefore obvious that the forest vegetation is contributing to the renovation of the percolating effluent; however, its order of magnitude is difficult to estimate because the annual storage of nutrients in the woody tissue and the extent of recycling of nutrients in the forest litter are extremely difficult to measure. Although considerable amounts of nutrients may be taken up by trees during the growing season, many of these nutrients are redeposited annually in leaf and needle litter rather than being hauled away, as in the case of harvested agronomic crops.

A comparison between the annual uptake of nutrients by an agronomic crop (silage corn) and a hardwood forest is given in Table 2.13. It is obvious that trees are not as efficient renovation agents as agronomic crops. Whereas harvesting a corn silage crop removed 145% of the nitrogen applied in the sewage effluent, the trees only removed 39%, most of which is returned to the soil by leaf fall. Similarly only 19% of the phosphorus applied in the sewage effluent is taken up by trees in comparison to 143% of the corn silage crop.

Table 2.13. Nutrient Uptake.
Annual uptake of nutrients by a silage corn crop and a hardwood forest irrigated with 5 cm of effluent weekly.

Nutrient	Corn Silage Pa. 602-A, kg/ha	Renovation Efficiency,[a] %	Hardwood Forest, kg/ha	Renovation Efficiency, %
N	180	145	94	39
P	47	143	9	19
K	144	130	29	22
Ca	30	15	25	9
Mg	26	27	6	4

[a]Percentage of the element applied in the sewage effluent that is utilized and removed by the vegetation.

WOOD FIBER QUALITY

The results of a recent study by Murphey *et al.*[7] provide some insight as to effects of municipal wastewater irrigation on the anatomical and physical properties of the wood of forest trees. Wood samples were collected from red pine and red oak trees irrigated with sewage effluent.

The report showed that sewage effluent irrigation on red pine resulted generally in increased specific gravity, increased tracheid diameter, decreased cell wall thickness, and no change in tracheid length. Likewise changes also occurred in red oak wood due to irrigation with sewage effluent. A 5% reduction in the early wood vessel segment diameter was also reported. These large barrel-shaped elements are the causes of "picking"–the lifting of the surface of paper during printing. The smaller, longer cells produced by effluent irrigation might reduce this problem when using ring porous woods such as red oak for pulp.

An increase in the number and height of broad rays resulted in an increase in the amount of wood volume occupied by the broad rays from 9% in the untreated xylem to 11.5% of the wood developed during irrigation. The increase in number and height of broad rays would cause an increase in the percentage of "fines" in a pulp mix. Increase in specific gravity and particularly in the change in the amount of latewood from about one-half to three-quarters of the growth ring provides more mass of fibers per unit volume.

Coupled with the growth rate change, irrigation with municipal wastewater results in the development of more fiber per treated tree. The increase in fiber and vessel segment length also increased the utility of this wood for pulp. Conclusions reached are that in general the alterations of the wood fibers resulting from wastewater irrigation enhanced their utilization as a raw material for pulp and paper.

ECOSYSTEM STABILITY

Ecosystems are somewhat elastic and can withstand a certain amount of stress prior to permanent change or collapse. Weekly application of wastewater will certainly impose a stress on the ecosystem. An unresolved question is whether the impact will be sufficient to cause a significant change and whether the change will be desirable or undesirable. Regular applications of large volumes of wastewater can turn a relatively dry site into a moist super-humid one and a relatively sterile site into a fertile one. Such changes may influence species composition and plant density on the site as well as fungi, bacteria, and microorganism types and populations. These changes, in turn, may influence the habitat and utilization of the site by wildlife.

In general these changes are subtle and occur over a long period of time. Since there are no municipal wastewater spray irrigation projects in the United States older than 12 years on which the ecosystems have been monitored annually, we can only speculate about the long-term effects. Results from the Penn State Project will illustrate some of the trends observed during the past 12 years.

Forest Reproduction

Forest reproduction was tallied on milacre plots in a mature mixed hardwood forest irrigated with sewage effluent at the rate of 2.5 cm per week during the growing season. Results indicated a drastic reduction in the number of tree seedlings present in the irrigated area. A pre-irrigation survey indicated about 39,000 tree seedlings per hectare in the control area and about 35,800 tree seedlings per hectare in the area to be irrigated. After 10 years of irrigation, remeasurement indicated only 4520 seedlings per hectare were present in the treated area. A similar reduction was also found in the number of herbaceous plants in the irrigated forest area. The initial survey indicated about 213,200 stems per hectare, whereas 10 years later there were only 36,500 stems per hectare.

Forest Floor

Measurements of the forest floor (leaf litter) made in the summer of 1974 indicate that drastic changes occurred in relation to the accumulation and decomposition of the forest floor. For instance, in the hardwood forest on the Morrison soil, which received the 5-cm per week application rate, the average depth of the forest floor in the irrigated area was 1.5 mm in comparison to 20 mm in the control area. Forest floor accumulations calculated on a dry-weight basis were 1206 kg/ha in the irrigated forest and 7566 in the control forest. Similar reductions in the amount of forest floor present were observed in all irrigated forest areas.

This accelerated decomposition of the forest floor may in time have detrimental effects on the physical properties of the soil. Without the protective leaf litter cover, the surface soil is more exposed to raindrop compaction which in turn might affect infiltration capacity. However, more important is the fact that the forest floor insulates the surface soil during the winter and prevents the formation of concrete frost. Without the thick forest floor cover, the exposed mineral soil is more susceptible to freezing, which might result in surface runoff and overland flow during winter irrigation periods.

Preliminary observations have also indicated that changes are occurring in the soil invertebrate populations (earthworms, mites, spring-tails) and in the infiltration capacity and percolation capacity of the 30-cm soil depth.

Old Field Herbaceous Vegetation

An Old Field area consisting primarily of poverty grass (*Danthonia spicata*), goldenrod (*Solidago sp.*) and dewberry (*Rubus flagellaris*) was irrigated with sewage effluent at the rate of 5 cm per week during the growing season since 1963. Significant changes have been observed in species composition, vegetation density, height growth, dry matter production, percentage areal cover, and nutrient utilization. Average dry matter production was 6111 kilograms per hectare on the irrigated plot and 2027 kilograms per hectare on the control plot. This represents an average annual increase of 201%. Annual increases ranged from 100 to 350%.

Several species that were predominant prior to wastewater irrigation have been drastically reduced in number or have disappeared completely. For instance, goldenrod (*Solidago sp.*), which had 383,000 stems per hectare in 1963, has been reduced to about 33,600 stems per hectare. White aster (*Aster pilosus*), which had about 303,700 stems per hectare in 1963, is no longer present on the site. The predominant species on the irrigated plot is clearweed (*Pilea pumila L.*), which covers more than 80% of the plot with approximately 47 million stems per hectare. This species is typical of shaded moist sites.

WILDLIFE HABITAT

Wildlife studies were initiated on the Penn State Project in 1971 and, hence, the results obtained to date are largely inconclusive. Wood *et al.*[8] reported that spray irrigation of sewage effluent at the rate of 5 cm per week in forests and on brushland appeared to have a favorable influence in the nutritive value of rabbit and deer forage. Foliar analysis of forage species indicated that crude protein, phosphorus, potassium, and magnesium contents all increased, whereas calcium content was decreased.

Studies using the lead deer technique to determine preference for or avoidance of irrigated areas indicated that the deer used treated sites as readily as untreated sites. During the winter period, observations indicated that wild deer used the irrigated areas for resting and feeding even though these areas were often ice covered as a result of winter irrigation.

No conclusive evidence was obtained on the effects of spray irrigation on rabbit reproduction. However, it was found that the winter carrying capacity of the irrigated sites exceeded that of the control areas presumably due to the higher levels of available nutrition and improved cover conditions.

Rabbits trapped on the irrigated areas appeared to be larger and healthier than those on the control area, which were as much as a third lighter in weight and showed obvious signs of emaciation.

Since public hunting areas and public gamelands are potential disposal areas, some concern has been expressed by sportsmen on the possible effects on game bird reproduction. It is well known that wet spring season periods may have an adverse effect on the survival of young broods of wild turkey and grouse. Many of the young chicks die of pneumonia during these cool, wet periods. An unanswered question is whether weekly spray irrigation of wastewater will produce the same condition.

TREE MORTALITY

In 1972 a survey was made in the hardwood forest area on the Morrison sandy loam soil to determine the effect of sewage effluent irrigation on mortality. This area was selected since it is the largest (approx. 8 hectares) forest stand under irrigation. Mortality was defined as all standing dead trees. The forest stand had received 10 cm of sewage effluent weekly during the growing season only from 1964 to 1967, during the dormant season only from 1968 to 1971, and 5 cm of effluent weekly during the growing season in 1972. The results of the survey are given in Table 2.14. Results indicated that mortality was about 20% higher in the effluent-irrigated forest area. There was also a large difference in the number of living trees per hectare, particularly, in the 5-cm diameter class. Unirrigated forest areas averaged 716 trees per hectare in the 5-cm diameter class compared to only 173 trees per hectare in the irrigated forest areas.

Table 2.14. Tree Mortality
Population and mortality in forested stands in control plots and irrigated with sewage effluent.

Diameter Class	Living Trees Stems per Hectare		Mortality Stems per Hectare	
cm	Irrigated	Control	Irrigated	Control
5	173	716	136	104
10	247	271	67	79
15	99	198	49	12
20	148	198	6	6
25	99	123	–	–
30	74	62	–	–
35	5	17	–	–
40	12	–	–	–
Total	857	1585	258	213

Although many of these young saplings are lost through natural suppression, a considerable number are lost in the irrigated areas through ice breakage during winter irrigation. The survey results indicated that approximately 128 stems per hectare showed visible ice damage in the irrigated areas in comparison to 18 stems per hectare in the control areas. Seventy-five per cent of these trees were in the 5-cm diameter class and the remainder in the 10-cm diameter class. The species most susceptible to ice damage was red maple. A reasonable amount of ice damage must be expected if disposal systems are to operate throughout the year. However, ice damage can be minimized through the proper design of the spray-irrigation system and the use of low-trajectory rotating sprinklers.[9]

A comparison of the results of the 1972 survey made in the effluent-irrigated forest areas with the results of several other forest surveys of mixed hardwood stands on a variety of sites in central Pennsylvania indicates that 10 years of effluent irrigation have produced no great differences. The average number of living trees, 13 cm in diameter and greater, in the effluent-irrigated forests was 437 stems per hectare in comparison to an average of 382 stems per hectare in several natural mixed hardwood stands. Mortality of trees, 13 cm in diameter and greater, in the effluent-irrigated forests was 55 stems per hectare in comparison to an average 52 stems per hectare in several natural mixed hardwood stands.

GROUNDWATER RECHARGE

The amount of renovated effluent recharged to the groundwater reservoir was estimated from data available on the total amount of effluent and rainfall received by the plots, and potential evapotranspiration. Annual recharge ranged from 10.3 to 17.3 thousand cubic meters per hectare irrigated with an average of 15 thousand cubic meters. Recharge amounted to approximately 95% of the effluent applied at the 5-cm per week rate. Recharge rates were higher when water was applied during years with normal or above normal rainfall throughout the year.

Evapotranspiration losses are greatly diminished during the late fall, winter, and spring, and more of the water infiltrating into the soil from natural precipitation and irrigation is potentially available for groundwater recharge. Runoff that did occur at the irrigation sites following snow and ice pack melt or heavy or prolonged rains was ponded in one or more closed surface depressions where it was captured by infiltration or it infiltrated in adjacent unirrigated buffer areas that were usually in forest cover. Closed depressions were numerous on all of the upland areas selected for irrigation or they were available downslope from test plots, hence detention storage was provided naturally. This would not necessarily be true at other irrigation sites where detention storage would have to be engineered to prevent or eliminate runoff. It was found that adjacent border areas with forest stands were ideal to help contain runoff and promote infiltration all seasons of the year. Rarely did overland flow extend more than 100 feet beyond irrigation plots during the spring thaw.

Mean annual concentrations of nitrate-nitrogen in deep groundwater monitoring wells on and adjacent to the spray irrigation areas are given in Table 2.15. Depths of these wells range from 100 to 300 feet. The 1962 values represent the pre-irrigation period. Spray irrigation of sewage effluent was initiated during the summer of 1963 near the F-wells and in 1965 near the G-wells. Background levels of nitrate-nitrogen concentrations can be inferred from the 1962 data and from the results of analyses of water samples collected from surrounding private wells.

Chemical water quality changes in the deep on-site wells have been nonsignificant at the cropland and forest areas with the Hublersburg soils (F wells). However, significant increases in nitrate and chloride have occurred in one well on the forested area with the Morrison sandy loam soil (G wells). Although there have been some increases in nitrate-nitrogen concentrations during the seven-year period (1963-69), these concentrations have remained well below the U.S. Public Health Service drinking water limits. Hence, it is evident that with properly programmed application, sewage effluent can be satisfactorily renovated and considerable amounts of high quality water recharged to the groundwater reservoir. In time, contributions to the groundwater of this magnitude will certainly have a beneficial effect on the local water table level.

CONCLUSIONS

Twelve years of research have indicated that the living filter system for renovation and conservation of municipal wastewater is feasible and that the combinations of agronomic and forested areas provide the greatest flexibility in operation. Such a system is more adaptable to small cities and suburbs than to large metropolitan areas because of the availability of open land close to the wastewater treatment plant, although the land area requirement is not a major prohibitive factor. At the recommended level of irrigation, 5 cm per week, only 52 hectares of land would be required to dispose of 4 million liters of wastewater per day. Although large contiguous blocks of agricultural and natural forest land would be the most desirable for efficiency and economy, major metropolitan areas could utilize golf courses, playing fields, forest preserves and parks, greenbelts, scenic parkways, and perhaps even divided highway and beltway medial strips.

Table 2.15. Groundwater Nitrogen.
Mean annual concentration of nitrate-nitrogen in deep groundwater monitoring wells on and adjacent to the spray irrigation areas.[10]

Well No.	Distance[a] in Meters	Mean Annual Concentration, mg NO_3-N/liter							
		1962	1963	1964	1965	1966	1967	1968	1969
G-3	0	0.2	–	–	0.1	–	3.6	3.5	2.8
F-1	30	–	1.4	0.7	0.4	0.9	0.9	0.5	0.4
F-5	60	–	–	–	0.4	0.6	0.4	0.8	0.3
G-10	150	–	–	–	–	–	0.9	0.3	0.4
F-3	360	–	–	–	0.4	0.6	0.4	0.5	0.4
UN-14	390	1.6	1.9	2.3	1.4	1.8	2.9	3.1	4.0
UN-24	660	0.6	0.6	0.3	0.5	1.2	1.0	0.9	1.3
UN-17	990	0.6	0.6	0.5	0.2	0.3	0.6	0.4	0.4
Private water wells in surrounding vicinity									
W-7	780	1.3	1.9	1.3	0.7	0.6	–	–	–
W-5	1590	5.3	5.0	4.0	3.0	3.2	3.7	4.4	5.7
W-30	1590	1.9	1.3	1.9	1.2	1.4	2.3	2.4	2.2
W-3	1710	10.1	9.2	4.6	3.2	5.0	7.6	10.2	7.3

[a]Approximate distance from sewage effluent application area

ACKNOWLEDGMENTS

Research reported here is part of the program of the Waste Water Renovation and Conservation Project of the Institute for Research on Land and Water Resources, and Hatch Project No. 1809 of the Agricultural Experiment Station, The Pennsylvania State University, University Park, Pennsylvania. Portions of this research were supported by funds from Demonstration Project Grant WPD 95-01 received initially from the Division of Water Supply and Pollution Control of the Department of Health, Education, and Welfare and subsequently from the Federal Water Pollution Control Administration, Department of the Interior. Partial support was also provided by the Office of Water Resources Research, USDI, as authorized under the Water Resources Research Act of 1964, Public Law 88-379 and by the Pinchot Institute for Environmental Forestry Research, Forest Service, USDA.

REFERENCES

1. Parizek, R. R., L. T. Kardos, W. E. Sopper, E. A. Myers, D. E. Davis, M. A. Farrell, and J. B. Nesbitt. "Waste Water Renovation and Conservation," Penn State Studies No. 23 (1967).
2. Sopper, W. E. "Effects of Trees and Forests in Neutralizing Waste," in *Trees and Forests in an Urbanizing Environment*, Coop. Ext. Service, Univ. of Mass. (1971), pp. 43-57.
3. Jackson, M. L. *Soil Chemical Analysis.* (Englewood Cliffs, New Jersey: Prentice-Hall, Inc., 1958).
4. Baker, D. E., G. W. Gorsline, C. B. Smith, W. I. Thomas, G. E. Grube, and J. L. Ragland. "Techniques of Rapid Analyses of Corn Leaves for Eleven Elements," *Agro. J.* **56**, 133 (1964).
5. Peech, M., L. T. Alexander, L. A. Dean, and J. F. Reed. *Method of Soil Analysis for Soil Fertility Investigations*, No. 757 (Washington, D.C.: United States Dept. of Agriculture, 1947).
6. Sopper, W. E. and L. T. Kardos. "Vegetation Responses to Irrigation with Municipal Wastewater," *Recycling Treated Municipal Wastewater and Sludge Through Forest and Cropland.* (Pennsylvania State University: The University Press, 1973), pp. 271-294.
7. Murphey, W. K., R. O. Bisbin, W. J. Young, and B. E. Cutter. "Anatomical and Physical Properties of Red Oak and Red Pine Irrigated with Municipal Wastewater," *Recycling Treated Municipal Wastewater and Sludge Through Forest and Cropland.* (Pennsylvania State University: The University Press, 1973), pp. 295-310.
8. Wood, G. W., D. W. Simpson, and R. L. Dressler. "Deer and Rabbit Response to the Spray Irrigation of Chlorinated Sewage Effluent on Wildland," *Recycling Treated Municipal Wastewater and Sludge Through Forest and Cropland.* (Pennsylvania State University: The University Press, 1973), pp. 311-323.

9. Myers, E. A. "Sprinkler Irrigation Systems: Design and Operation Criteria," *Recycling Treated Municipal Wastewater and Sludge Through Forest and Cropland.* (Pennsylvania State University: The University Press, 1973), pp. 324-333.
10. Kardos, L. T., W. E. Sopper, E. A. Myers, R. R. Parizek, and J. B. Nesbitt. *Renovation of Secondary Effluent for Reuse as a Water Resource*, EPA 660/2-74-016 (Washington, D. C.: Environmental Protection Agency, 1974).

3

TREATMENT PROCESSES AND ENVIRONMENTAL IMPACTS OF WASTE EFFLUENT DISPOSAL ON LAND

Demetrios E. Spyridakis and Eugene B. Welch

Associate Professor and Professor
Department of Civil Engineering
University of Washington
Seattle, Washington 98195

INTRODUCTION

The rationale underlying this study derives from general and specific needs in the area of utilization of the earth's soil mantle as an engineering system for wastewater disposal. Utilization of the soil as a "living filter" has been acclaimed as the most ecologically appropriate treatment technique because recycling of materials is a built-in reality.[1] However, the process of wastewater purification and treatment on land is very complex because of the large number of interacting variables involved. Since time immemorial, land disposal has occurred on a natural scale and it has produced both beneficial results and environmental disasters. Man-made wastes have been added to the land for centuries. Within the boundaries of a certain ecological milieu such additions often proved to be highly efficient, particularly when practiced on a scale involving small sources or small regions.[2] Today there is undeniable evidence that skillful land disposal augments soil fertility and provides unique means of energy preservation. On the other hand, promiscuous use of borrowed methods and wastes in unsuitable environments has yielded disappointing results. Knowledge about the implications of large scale use of land disposal is presently insufficient and warrants additional scientific information concerning waste component transfer in the living environment.

This chapter concentrates principally on the degree of wastewater renovation as related to a particular soil under given geologic and climatic conditions. Three basic approaches to land application are considered: spray irrigation, overland runoff and rapid infiltration ponds. The extent of renovation of chlorinated domestic-industrial secondary effluent and potential environmental impacts of renovated wastewater are estimated for model soil types and a climate characteristic of the southern Great Lakes.

Some well-known principles are reviewed briefly and some properties of the soil-water-plant system will be discussed concerning (1) reactions that effluent constituents undergo in land disposal systems and (2) factors and processes influencing the amounts and kinds of nutrients and/or contaminants leached from the application site. The emphasis will be primarily on Biochemical Oxygen Demand (BOD), nitrogen, phosphorus, trace metals, organic toxicants, viruses, and bacteria. The information and rationale employed in selecting land disposal sites and amounts of application and predicting degree of renovation will be presented. Hazards that may be associated with the sensitivity of a given plant species to certain levels of toxic constituents are not considered. Also this chapter does not deal with those aspects of land disposal associated with the correlative long term dynamic changes that will undoubtedly occur in the product quality and in the composition of soil under the influence of excessive fertilization, altered growth patterns, reduced removal of transpirational water from the deep soil layers, and modifications in the site's microclimate, fauna, and flora.

RENOVATION FACTORS AND PROCESSES

Pollutant transformations, and translocation rates of the pollutants and their degradation products in soil are a function of the combined effects[2-8] of chemical, microbial and physical soil parameters; these parameters must be expressed in an integrated form as measurable soil properties to achieve satisfactory prediction and control of the soil phase of pollutant cycles. The fate of wastewater materials in soils is determined by a large number of processes, including physical retention, adsorption on solid surfaces, plant and microbial uptake, microbial degradation, volatilization, leaching, chemical breakdown, and precipitation. The relative importance of each of these individual mechanisms in wastewater renovation in soils will ultimately be a function of concentration of the material and soil properties.

The following section briefly reviews the factors and mechanisms that may contribute to wastewater renovation in land disposal systems. The major transport agent of pollutants in soil is water. Pollutants are translocated down through the soil profile into groundwater and are carried, often closely associated with soil particles, in surface runoff. Soil particles carried into surface waters by runoff become an intimate part of the sediment component of the water systems and may cause contamination or decontamination of the waters depending on the relative pollutant contents of the soil (sediment) particles and the waters.

The soil factors that determine effectiveness of wastewater renovation are: soil type as determined among other parameters by the soil's content in clay minerals, fine particles and organic matter, soil permeability, soil pH, groundwater level, slope of ground surface, nearness to a stream or other surface body of water, depth of soil column, type of bedrock material, and vegetation cover. The soil's ability to purify wastewater depends, in addition to soil characteristics, upon design, type, configuration and loading of the application system, and maintenance of the land disposal system.[2,6,9]

The soil factors that primarily determine the concentration of a given nutrient or pollutant in soil solution are, in relative order of importance: (1) secondary minerals that are, in general, confined to the finer fractions smaller than 2μ; (2) organic matter, combined with aluminum, iron or other metals; (3) uncombined oxides, carbonates and sulfates that may contain calcium, magnesium, iron, aluminum or manganese; and (4) primary minerals confined to the coarser fractions of the soil that are greater than 2μ. Uncombined oxides and carbonates exist in both fractions, both as precipitates, particularly in the larger fractions, and as coatings on both primary and secondary minerals. The organic matter may exist in all stages of decomposition, in all size distributions, and both as entities and as coatings or even in chemical combinations with the mineral phase.

The interactions between the solid phase and the soil solution comprise several forms of binding. Part of these are electrostatic and are influenced primarily by the valence of the ion and the charge density of the solid phase. For cations the interaction is mainly governed by charge effects; for anions the interaction is somewhat different and chemical binding is often dominant.

The charge of the three-layer type minerals is due mainly to isomorphous substitution, which results in a constant negative charge density on the surface of the crystal. In addition to isomorphous substitution, ionized -SiOH and –AlOH groups of two-layer clay minerals and –FeOH or –AlOH from coatings may sorb both cations and anions, depending on the pH of soil solution. Thus it is apparent that many of the exchange properties of the soil are due to contributions from oxide coatings.

The organic matter also provides a reactive surface that both sorbs cations in exchangeable positions formed by COOH and OH groups and also may complex such ions as Fe, Mn, and even Ca and Mg, which in turn may adsorb anions such as phosphate.[10] In very sandy soils, organic matter becomes the dominant factor determining the concentration of many nutrients and pollutants in the soil solution. The cation exchange capacity of the humic acid, principal component of soil organic matter, approximates 250-400 me/100 g, which is three times that of the montmorillonite-type secondary clays and 30-100 times that of the kaolinite type.

The soil factors and properties described thus far, indispensable as they are for the proper function of soil as a medium of life sustenance, cannot alone provide the desired effluent renovation. It is the life that permeates every minute fraction of soil that will provide ultimately the unique mechanism of effluent recycling and renovation. The most important living constituent of soils is the root system of vegetation with the associated fungi and bacteria that assist the plants and trees in uptake of nutrients. The population of organisms varies from a small community of anaerobic and acid-tolerant microbes to a multitude of bacteria, actinomycetes, fungi, algae, protozoa, nematodes, worms, arthropods, moles, shrews, and rodents.[11] Their individual functions and contributions on land wastewater treatment are beyond the subject matter of this chapter. It suffices to state that their role is manifested in the fate of all effluent components whether it be loss of nitrogen to the atmosphere through denitrification, immobilization and deactivation of toxicants or degradation and recycling of added organic loads to the land.[12,13]

MICROORGANISM REMOVAL FROM WASTEWATER

The removal of microorganisms, particularly human pathogens, from wastewater as it contacts the soil is an important consideration in land disposal of domestic waste. The residual concentration of microorganisms in treated wastewater is low; however, the efficiency and continuity of disinfection determine the magnitude. The greater resistance of viruses to routine disinfection procedure and the possibility of chlorination failure and disposal of wastes not disinfected increases the importance of the capacity of the soil itself to remove organisms. Once removed the longevity of pathogens in the soil becomes of interest. Potential for local contamination by air transport should also be considered.

Extensive field observations indicate that bacteria and viruses are efficiently removed from wastewater as it percolates through soil. Viruses are probably transported to greater depths in some instances than bacteria because of their smaller size;[14] however, virus transport is also considered minimal and some believe at no greater rate than bacteria.[15] "Gravity and time are against bacterial contamination of an outcropping ground water," and unless fissures or dissolution channels are present for organism transport, percolation through even the coarsest soil will remove bacteria and viruses within a few to several feet.[15,16]

Although several feet of soil appear necessary for nearly complete removal of bacteria, most of the organisms are removed in a shallow film of soil at the surface (upper 1 cm). Observations show that as much as 92-97% removal occurs in this upper 1 cm layer.[17] These results are particularly significant in considering overland runoff as a treatment method in areas with soil of high clay content and low permeability. In such situations percolation may occur in the first few centimeters only.[18] Percolation may be minimal but since wastewater flows continually over and through a few hundred feet of this effective top centimeter, removal efficiency with overland runoff may be almost as effective as vertical seepage. This may be particularly true if other mechanisms besides physical filtration are considered, *e.g.,* biological competition.

Removal efficiency (per cent removed per distance travelled) from wastewater disposed in rapid infiltration ponds is probably less than either of the other two methods of disposal. However, distance travelled and detention time is greater, so ultimate removal may be as great as with spray irrigation. Field observations indicate efficient organism removal under high rates of infiltration. At Santee, California, treated wastewater was percolated through 1500-foot (457 m) channels composed of sand and gravel. The flow rate through these channels was about 100 feet (30 m)/day, which represents a very rapid infiltration rate through a porous medium. Most of the bacteria in the effluent, and all viruses that were experimentally added, were removed in the first 200 feet (61 m).[19] At Flushing Meadows near Phoenix, Arizona, wastewater is added to infiltration basins consisting of 3 ft (1 m) of fine loamy sand underlaid by a succession of coarse sand and gravel layers to a depth of 250 feet (76 m). This upper layer contains only 2% clay. The infiltration rate was 330 feet (100.6 m) of wastewater per year or about one foot (0.31 m) per day. At about 30 feet (9.1 m) from the point of infiltration [8 feet (2.4 m) vertically and 20 feet (6.1 m) horizontally] total coliforms had decreased to a level of 200/100 ml during 2-3 week inundation periods and to 5/100 ml during 2-3 day periods, a removal of more than 99.9%. Fecal coliforms in this water were usually near zero. An underground detention time of about one month was considered adequate for complete removal of all coliforms.[20]

The groundwater flow and hence detention time of the treated wastewater is of considerable importance to reuse of the reclaimed water for drinking. Organisms are not removed as rapidly (per cent removal per distance) from horizontally as from vertically flowing groundwater in unsaturated soil. Organisms reaching the groundwater may travel more than 25 feet (7.6 m) horizontally with a groundwater flow of 25 feet (7.6 m) per day.[21] Coliform organisms as well as enterococci travelled no farther than 100 feet from an aquifer recharge well following addition of varying amounts of primary sewage.[22] Thus, travel distances of 150-200 feet (45.7-60.9 m) should result in adequate organism removal from wastewater by rapid infiltration. A detention time of several days over this distance seems advisable before reuse is permitted.

The efficiency of microorganism removal from wastewater applied to soil would be expected to be great under either spray irrigation, overland runoff or rapid infiltration. Probably the least efficient technique would be overland runoff, although even here filtering capacity of particulates by the plant debris on the surface is fairly efficient. Rapid infiltration ponds would probably attain an ultimate removal capacity similar to that of spray irrigation, but as mentioned earlier, it will require a greater distance and detention time.

BIOLOGICAL MECHANISMS OF ORGANISM REMOVAL

Organisms in wastewater applied to soil are faced with competition for space, food supply and antibiotic materials from other microorganisms and also predation by larger soil organisms. Competition is probably greatest in this shallow surface layer since oxygen is more abundant and rates of decomposition are greater, which would lead to greater food supplies for larger organisms. Thus, human pathogens not adapted to the rigors of such an existence do not survive long. This has been demonstrated to some extent in that *E. coli* survival was considerably greater in sterile soil than in nonsterile soil.[23] McGauhey and Krone[24] cite other work that shows low survival of organisms from human intestinal origin as a result of competition with normal soil microflora. The extent to which biological competition and simple physical filtration contribute to microorganism removal from wastewater in each case is not known. Since removal efficiency seems to be high in nearly all soil types examined, and no great variation occurs with seasons and temperature, one might conclude that physical filtration is the principal removal mechanism. The persistence of these organisms once removed by the soil depends on biological controls regulated by environmental factors such as temperature; in some instances survival in soil of organisms from intestinal origin has been relatively great.

The concentration of organisms in wastewater applied would be the tolerant residual following normal disinfection. Coliform bacteria (total and fecal) are used as indicators of disinfection effectiveness on a routine basis because pathogens, and particularly viruses, are difficult to measure. Although the effectiveness of pathogen removal, particularly viruses, and the validity of the coliform index are controversial subjects, it appears that normal disinfection practices are reasonably effective at removal of pathogens (including viruses) from treated sewage effluent. Under routine operation Chambers[25] reported results of coliform survival from 45 treatment plants. Most (93%) of the plants showed that less than 5000/100 ml survived disinfection, and over one-half (56%) showed that less than 500/100 ml survived. This represents at least 99.9% removal in most plants.

The longevity of such a small density of residual bacteria in the soil varies depending upon environmental conditions. Temperature, organic matter content and whether or not the soil is aerobic are the more important variables. The *Salmonella* typhoid bacillus has been observed to survive in soil from six months to a full year in some instances[23] and a maximum of about one month in others.[24] Survival increased with increased organic content in the soil; survival decreased to less than one week when sand was used.[24]

Coliforms survive for four years in soil,[26] longer on the average than typhoid or tuberculin bacilli.[23] Moisture was a significant factor, with survival in wet soil double that in dry.[23] As stated previously, biological competition and predation are important determinants of enteric organism longevity in the soil; consequently survival can be expected to be greatest when normal biological activity is least, *i.e.,* low temperature and anaerobiosis. Thus, findings that *Salmonella* survive two years in frozen soil, but a maximum of one year in unfrozen soil and that *E. coli* survive longer in sterile than in nonsterile soil are not surprising.[23]

Although longevity of pathogens could be significant during the beginning and ending of the spray period in cold climates, survival would not usually be expected to exceed one month if biological activity is relatively great. If death and decomposition rates and loading rates of pathogens are constant, significant accumulation of pathogens would also not be expected even though longevity can be a month or more.

Air dissemination of surviving pathogens from wastewater irrigation sites should be considered as a possible hazard even though the likelihood of disease spreading from such sites would at first seem remote. Spray irrigation would probably expose a small percentage of the particles, including bacteria, in the wastewater to action of the wind. This presumably minimal opportunity for wind removal of particles from sprayed droplets together with the very low probable concentration of pathogens in the wastewater suggests that the hazard of airborne transmission of disease from spray fields (if wastewater is disinfected) would be slight. Measurement of air dissemination of enteric organisms from sites of land disposal of wastewater are limited, but in one instance coliforms were detected up to 200 feet (61 m) from the point of application.[27] However, extra safe precautions would seem to be in order, and residential areas should be kept at a considerable distance from the spray fields even though dilution is great and survival poor.

BOD REMOVAL

BOD is removed from wastewater in the soil mantle by a combination of physical and biological processes. The capacity of the soil and its microflora as a physiobiological remover of BOD is enormous. This capacity can be exceeded, however, by adding a greater mass of BOD than can be decomposed within the time that the water is detained in the soil or by applying more wastewater than the soil will absorb (hydraulic overloading). In both cases anaerobic conditions may result, which would greatly reduce the decomposition rate and increase the possibility of clogging. Clogging can result from various processes, but the effect of biological activity seems to be the most important.[24]

The BOD removal efficiency of soil can be affected by the amount of vegetation cover and the infiltration capacity. Anything that adds surface area at the soil-air interface, whether it be litter or living plants, will increase biological decomposition capacity. At very high rates of wastewater infiltration, residence time of the dissolved or particulate BOD may not be great enough for complete biological decomposition to occur and a sizable fraction of undecomposed BOD may reach the groundwater. However, net BOD removal efficiency has been observed to be quite high even in the coarsest soils and highest infiltration rates studied, even though the removal efficiency per unit soil column decreased under such conditions. The soil acts as an effective filter in removing particulate and dissolved organic matter; most of this removal occurs in the upper 5-6 inches (12.7-15.2 cm) in the profile.[24] This physical removal can account for as much as 30-40% of BOD, COD (chemical oxygen demand) and TOC (total organic carbon).[18]

The particulate organic matter that is filtered by the soil as well as that dissolved in the percolating water is partially degraded by microorganisms. Soils contain a large complement of heterotrophic microorganisms that allow the total system the ability to utilize and degrade a broad array of organic compounds under a variety of environmental conditions. Most materials eventually degrade. The time required, however, may range from minutes for glucose, to hundreds of years for a complex aggregate of compounds called soil humus.[28] Under aerobic conditions the bulk of the degraded mass will evolve from the system as CO_2. If the active surface layer of the soil is anaerobic, degradation will stop short of complete conversion to CO_2 and many reduced organic compounds result from the incomplete degradation. Under both conditions, mineralized inorganic materials are formed and either taken up by plants, held in the soil, or transferred through the soil mantle with the water flow. The degree to which these are removed from the soil and enter the groundwater determines, in part, the environmental effect of the wastewater. A portion of the degraded organic matter is reconstituted into biological cells (sludge), a substrate for further microbial decomposition or clogging problems in the soil mantle, which contributes to food for larger animals such as insects and annelids (worms).

The nature and percentage of organic matter in treated sewage effluent that is included in its BOD is of interest because it is the relatively refractory materials that are most apt to resist biological removal in the soil mantle and may reach the groundwater. Although little is known of the kinds and concentrations of organic compounds in secondary waste,[24] some work shows that only about 30% of the organic carbon in secondary treatment effluent could be assigned to chemical groups.[29,30] Whereas 50% of the soluble organics in settled sewage are carbohydrates, only 10% could be ascribed to this category following secondary treatment. Most compounds after that stage are considered to be soluble acids. Thus, the immediate carbonaceous BOD in secondary effluent is probably a small fraction of the total dissolved organic matter. Simple sugars, starches, hemicelluloses, celluloses and proteins apparently decompose relatively rapidly in soil, whereas such groups as lignins, waxes, tannins, cutins and fats are more resistant to decomposition.[24] If anaerobic conditions are created by too great a hydraulic or BOD loading for the soil capacity, then these more resistant compounds are apt to accumulate.

In the light of the above discussion it is reasonable to conclude that a model secondary effluent with a BOD concentration of 25 mg/l and an application rate of two inches (5.1 cm) per week or 11.5 lb/ac wk (12.8 kg/ha wk) would not result in appreciable accumulation of undecomposed organic matter in the soil mantle. Since soils have remained aerobic under loading rates of hundreds of pounds per acre per day with percentage removals in the upper 90's, it would appear that treatment of model wastewater such as the above would be limited more by hydraulic capacity of the soil than by the ability of its microflora to remove the BOD. Since disposal sites are normally selected to accommodate hydraulic loading of at least two inches (5 cm) per week, BOD removal under either spray irrigation or rapid infiltration can be expected to be highly efficient. Efficiency may be less with overland runoff. The following literature review tends to support this contention.

Industrial wastewater containing 1,150 mg/l BOD_5 was sprayed on sand with high permeability and well covered with reed canarygrass and humus.[31] BOD removal was reported greater than 99% with a BOD loading of 138 lb/ac day (155 kg/ha day). Based on preliminary experiments, it could be concluded that effective operation of the system depended upon hydraulic loading. At 3 inches (7.6 cm) per day the soil remained too wet and the canarygrass, very important to the soil's percolation capacity, was killed.

Thus, treatability of a large volume waste with low BOD may be limited by percolation capacity of the soil, while a small volume waste with high BOD is more apt to be limited by the oxidative capacity of the microorganisms and sorptive capacity of litter on the soil surface.

During eight months of the year paper mill wastewater was sprayed onto very permeable Norfolk-type sand at the rate of 3.6 in (9.1 cm)/week. This represents a loading rate of 181 lb/ac-wk (203 kg/ha-wk), at a BOD concentration of 219 mg/l. Although from 3,889 to 6,340 lb of BOD were applied per acre (1,764 to 2,876 kg/ha) during this period, no change in the organic matter in the soil was observed.[31]

Several examples of wastewater treatment by application to soil show that BOD removal efficiency is high even if loading is relatively great (Table 3.1). The consistently high removal efficiency of BOD (⩾95%) from wastewater applied to soils by spray irrigation is evident for a variety of soil types and application rates, as shown in Table 3.1. Only in very coarse soil with exceptionally high infiltration rates would the physiobiological mechanism fail to remove most of the BOD within a shallow surface layer of soil. Even with rapid infiltration BOD removal is ultimately very efficient. At Flushing Meadows, BOD in secondary plant effluent is reduced from 15 to 0.3 mg/l (98%) over a combined vertical and horizontal distance of about 30 feet (9.1 m).[20] The application rate here ranges from 1 to 4 feet (0.31-1.2 m) per day.

Table 3.1. Summary of Data from Sites of Wastewater Treatment by Soil Application[a]

Activity	BOD mg/l	Application Rate in./wk[c]	BOD Loading lb/ac-wk[d]	Soil Type	Removal Efficiency %
Seabrook Farms Food Process	1000[b]	14	3220	loamy-sand	98
Riegal Paper Co. Industrial	600	3.85	531	sandy	95
Nat'l Fruit Products Food Products	2600	8.75	5232	sandy-loam	99
Bearmore & Co. Industrial	600	8.4	1159	glacial till	95

[a]Data of Elazar.[32]
[b]Estimated considering usual high BOD of food process waste.
[c]To convert in./wk to cm/wk multiply by 2.54.
[d]To convert lb/ac-wk to kg/ha-wk multiply by 1.12.

Overland runoff is expected to provide a less efficient removal of BOD than either spray irrigation or rapid infiltration. Wastewater from the Campbell Soup Co. plant at Paris, Texas, was applied to four experimental watersheds with rather impermeable soil, and treatment was accomplished by overland runoff.[18] The average BOD in the wastewater was 572 mg/l. The application rate was 3.08 inches (7.8 cm) per week with a resulting BOD loading of 405 lb/ac wk (454 kg/ha wk). Very high removal efficiencies (>98%) were observed in the system even though percolation into the soil below the first inch (2.5 cm) or so was slight [percolation rates of 0.12-0.17 in. (0.31-0.43 cm)/day]. The high efficiency at this site is partly a function of the high BOD (572 mg/l) applied. Probably because of less filtering capacity by this method, a fairly high residual BOD remained (mean 9 mg/l).

Soil clogging associated with excessive BOD loadings can severely limit the site's renovation function; although it would not be expected on well-managed disposal sites at reasonable application rates such as two inches (5.1 cm) per week, the conditions that would cause such a problem should be considered in order to understand better the capacity of soil for wastewater treatment. Clogging can result from biochemical reactions, excessive loading of inorganic and organic materials (both particulate and dissolved), excessive hydraulic loading, and geometry of the soil surface and profile. However, the most significant cause is the activity of microorganisms as it pertains to the conversion of dissolved organic matter into biological sludge (cells) as well as to the failure of microorganisms to decompose filtered organic particulate matter.

Clogging usually occurs in the top few inches of soil[33,34] and is more a function of the organic mat that is largely independent of the coarseness of the soil.[24] The perpetuation of anaerobic conditions in the soil surface layer leads to clogging. Anaerobic conditions result in a low rate of biological activity and, thus, a tendency for sludge accumulation, production of ferrous sulfide and/or accumulation of polysaccharides. The anaerobic conditions could be brought about by excessive BOD or hydraulic loading, but usually it is hydraulic overloading that leads to inability of the soil to accept the water application rate, and ponding results. The "off-on" application procedure allows for drying and reestablishment of aerobic conditions in the soil. If ponding persists, clogging problems increase. Reduced moisture contents are necessary to attain maximum decomposition and minimum humus buildup.

Experimental columns with sand and gravel clogged three to ten times faster under anaerobic than under aerobic conditions.[33] Continuous ponding of applied septic tank effluent under anaerobic conditions decreased conductivity through the columns. Laak[34] also observed that domestic wastewater (untreated) rapidly clogged soil in experimental columns. Clogging was attributed to bacterial cell (sludge) formation in response to BOD consumption. Bacterial cells comprised 90% of the material that completely clogged the soil surface layer after 180 days.

Allison demonstrated that soil clogging is a result of biochemical processes by organisms within the soil and not a result of filling soil spaces with organisms added with the wastewater.[35] He found that continuous inundation of the soil resulted in clogging and was therefore a result of internal processes and obviously not caused by added organisms. A typical biologically-clogged soil may appear as a heavy overgrowth of black biological slimes.[24] This slime layer is not all bacterial cells but, to a large extent, comprised of ferrous sulfide. However, FeS was found to penetrate to considerably greater depths than the biological mat.[36] Since the area of clogging is generally confined to a few inches at the surface, it was suggested that the biological mat was relatively more important in clogging than anaerobically formed ferrous sulfide.

Polysaccharides and polyuronides are also heavily implicated as being responsible for soil clogging. The accumulation of these compounds in the soil is inversely related to the infiltration capacity. These compounds, produced by microorganisms utilizing readily decomposable organic matter added to the soil, accumulate under anaerobic conditions but are rapidly broken down by aerobic organisms when the soil is aerated.[37] According to McGauhey and Krone,[24] the process of ferrous sulfide and polysaccharide accumulation helps to explain why continual loading of soil leads to clogging. But by employing resting stages for the soil to dry, the clogging problem disappears. This is apparently caused by the column of air, which, drawn into the soil to replace the water, changes anaerobic to aerobic conditions.[38] Although temperature was not found to be particularly critical for the success of polysaccharide-consuming organisms, pH was very critical and must be maintained above 7.

From the preceding evidence, it becomes clear that conditions of water ponding and continual addition of wastewater must be avoided. Since clogging problems were apparently not observed at the various sites that received exceptionally high BOD loading rates, it is unlikely that moderate BOD loading of secondary effluent contributes enough substrate to produce clogging problems provided that aerobic conditions are usually maintained. As the evidence indicates, hydraulic overloading by itself can produce anaerobic conditions and clogging as a result of biological processes within the soil, regardless of BOD or organisms added.

REMOVAL OF PHOSPHORUS AND NITROGEN

Plant uptake of phosphorus and nitrogen combined with biological, chemical and/or physical immobilization of phosphorus, volatilization of nitrogen through biological and/or chemical denitrification, escape of NH_3 from alkaline soils, and NO_3 leaching provide the main mechanisms for nitrogen and phosphorus removals from soil solution.

Phosphorus as orthophosphate reacts with practically all soils with an almost quantitative removal from solution.[39,40,41] This is not surprising as soils characteristically have very reactive surfaces containing iron, aluminum, calcium, all of which form very insoluble phosphates. Acidic conditions favor the Fe-P and Al-P and alkaline conditions favor the Ca-P retention.[39,42] Retention of organic phosphorus compounds at colloidal surfaces may also take place and alter their chemical and biochemical stability. The mechanisms of phosphorus retention are controversial. The retention of phosphorus by aluminum and iron apparently involves both precipitation (reaction with iron and aluminum cations in solution) and adsorption (reaction at the surfaces of iron and aluminum compounds) where adsorption predominates at low phosphorus concentrations. Mechanisms of phosphorus retention by calcium may be similar, which involve the formation of more insoluble compounds such as $Ca_4OH(PO_4)_3 \cdot H_2O$ and $Ca_{10}F_2(PO_4)_6$ or adsorption and/or occlusion of phosphorus onto $CaCO_3$ precipitates.

Certain reactions may lead to the production of nitrogen gases and thereby constitute a possible pathway for nitrogen losses from soils. Both inorganic and biological reactions may lead to the volatilization of nitrogen gases. However, the contribution of chemical denitrification to soil nitrogen losses is not well documented.[39,43,44] The inorganic reactions include decomposition of HNO_2 under acidic conditions to give NO, which may be partially converted to NO_2 and HNO_3, and Van Slyke type reactions of HNO_2 with amino acids or NH_4 to yield N_2. Such reactions may partially account for nitrogen losses attributable to biochemical denitrification.[10]

The crop itself can remove certain amounts of nitrogen and phosphorus and incorporate them in the biomass (Table 3.2). In normal agricultural practices, the nitrogen and phosphorus recovery from fertilizers in crops are both about 50% and 30%, respectively. Only with very careful application of fertilizers can these removals increase to 70-80% and 50% for nitrogen and phosphorus, respectively.[10] However, in one irrigation field in the Pennsylvania studies, nitrogen uptake by the vegetation of more than 100% was calculated under a load of 2 inches (5.2 cm) per week.[51-54] Another important impact of vegetation is its direct synergistic effect on the denitrification process. Oxygen uptake by the roots that can cause locally anaerobic conditions and the decrease in oxygen diffusion due to the thick

Table 3.2. Annual Uptake of Elements by Forest and Crops in Relation to Quantities Applied with Wastewater[a]

	Wastewater[b]	Forest			Crops		
Element	lb/ac	lb/ac	%[c]	Reference	lb/ac	%[c]	Reference
N	351	20-80	14	46-49	40-300	48	2,6,10,39
P	176	2-20	6	46,49	2-45	13	2,6
K	246	3-65	14	46,49	50-300	71	2,6
Ca	421	6-50	7	46,49	10-150	19	2,6
Mg	298	3-8	2	50	2-50	9	2,6
Na	878	1-3	0.2	50	2-40	2	2,6
Fe	9	0.1-0.5	3	50	<1	11	10
Mn	9	0.5-1	8	50	<1	11	10
Cu	4	<0.05	<1	50	<0.1	<3	10
Zn	5	<0.1	<2	50	<0.1	<2	10

[a]To convert lb/acre to kg/ha, multiply by 1.12.
[b]These figures are based on an average composition of effluent and on the assumption that 78 acre-inch (3250 m^3/ha) of effluent are applied per year.
[c]Per cent of total applied.

vegetation cover may promote denitrification. Moreover, vegetation decreases soil erosion under both spray irrigation and overland runoff applications of wastewater with subsequent reduction in the transportation of nitrogen and phosphorus associated with soil particles to surface waters through surface runoff.

The efficiency of nitrogen and phosphorus removals obtained with different land disposal systems of wastewater on a variety of soils is summarized in Table 3.3. In general nitrogen removal varies from less than 0%[55,56] to 91%[52-54] removal depending primarily on soil type and depth of soil column, design and mode of application of wastewater and vegetation cover. The phosphorus removal varies from 0%[52,56] to 99%[51,56,57] and is governed primarily by soil type and depth of soil column. The specific effects of the most important operating factors in land disposal systems are briefly discussed hereafter.

The phosphorus removal increases with the clay content of the soil or decrease of the permeability.[63,64] Generally phosphorus is effectively removed in the upper 1 to 2 feet (0.31-0.62 m) of the soil column by adsorption/precipitation reactions when clay, sesquioxides of iron and aluminum, and calcareous materials are present. It is estimated that every

Table 3.3. Nitrogen and Phosphorus Renovation in Relation to Land Disposal Systems and Soil Conditions[a]

Amount Sprayed in./day[b]	Dose-Schedule	Duration	Renovation, % N	Renovation, % P	Soil Type	Soil Cover	References
			Rapid Infiltration Ponds				
36	2 day infiltration 2 day rest	year round	0-10	62	silty sand	bare-grass	20,55
36	14 day infiltration 14 day rest	year round	30	62	silty sand	grass	20,55
55	6 month infiltration 6 month rest	183 days	92	93	sand	grass	60,61
6	7 day infiltration 7 day rest	year round	11	100	sandy loam	bare	57
6	14 day infiltration 7 day rest	year round	42	94	sandy loam	bare	
6	continuous 120 days few months rest		86	55	sandy loam	bare	57
12	1 dose/day 10 hours	year round	0-58	0-34	sand	bare	56
			Overland Runoff				
0.9	2 doses of 6 hrs 5 days/week	150 days	90	65-84	clay loam	grass	60
0.63	1 dose of 7 hrs 5 days/week	year round	84	56	sandy loam	grass	61,18
0.63	1 dose of 7 hrs 3 days/week	year round	84	83	clay loam	grass	61,18
			Spray Irrigation				
2	1 day/week	Apr.-Nov.	80	99	loam	forest	51-54
1.4-2	2-3 days/week	Apr.-Nov.	91	99	loam	forest	51-54
2	1 day/week	52 weeks	0	91	loam	forest	51-54
4.5	3 day sprayed 4 day rest	year round	56	91	silty sand	forest	63
2	1 day/week	year round	97	99	loam	reed canarygrass	54

[a]Renovation is expressed as percentage of the original effluent element content that is retained in the soil utilized by vegetation or volatilized. Concentrations of N and P in the effluents used in the experiments reported in the table were around 20 and 10 mg/l, N and P, respectively.

[b]To convert in./day to cm/day multiply by 2.54.

10 years a depth of 1 foot (0.31 m) will be saturated with phosphorus, which in the long run can limit the use of the soil. When spray irrigation runoff is practiced this may become a limiting factor over longer periods of time because runoff water does not penetrate very deeply.

Nitrogen removal is markedly influenced by the soil type. An increasing clay content favors denitrification due to its effect on the soil structure,[51,60,62,65,66] Well-aggregated soils offer an ideal environment for biological denitrification through a combination of aerobic-anaerobic conditions in the microenvironment of aggregates. For example, on the perimeter of an aggregate, aerobic conditions are present while inside the aggregate the NO_3 under low oxygen pressures can be reduced to N_2 or N-oxides. Indications that such a system may be operating in soils is provided by more recent findings[43-45,67,68] indicating that soil nitrogen losses through biological denitrification may be much more appreciable than originally estimated. Nitrogen losses as high as 120-320 kg N/ha have been reported for rice soils with midseason drying.[44] This mechanism of nitrogen removal offers the only means by which NO_3 leaching to groundwater (a severe limitation of land disposal) can be partially controlled. Indispensable to nitrification-denitrification mechanisms in a land disposal system is a soil column depth of at least 2 feet (0.62 m) where the two reactions can occur simultaneously.

The effect of the dosage, schedule of application, and amount of wastewater applied (hydraulic loading) can be described in the same way as that of a normal sewage treatment system. A higher load results in a larger mass removal and a lower removal percentage. Increased loadings under certain conditions may shorten the longevity of the soil due to organic matter accumulation that cannot be degraded rapidly enough. Higher moisture contents that are the result of higher loadings create locally anaerobic conditions that are necessary for denitrification. Normally nitrogen removal increases with loading until this reaches a point where NH_4^+ cannot be effectively nitrified any more and nitrogen removal drops sharply.[67,69] On the other hand, a single dose per week creates highly aerobic conditions in the soil that do not favor denitrification. At four doses per week, denitrification may remove 60 to 90% of the applied nitrogen.[55] A higher dosage per week also increases carbon removal but may shorten the longevity under high loading conditions. A selected dosage will therefore always be a compromise between nitrogen and carbon removal and longevity. A 2-inch (5.1-cm)/week load of the secondary wastewater effluent considered in this report should not adversely affect the longevity of the soil.

The nitrogen removal in rapid infiltration ponds is governed mainly by the period the basin is inundated.[20,55,65,70,71] A longer period gives a better balance between aerobic and anaerobic conditions necessary for nitrification-denitrification. However, prolonged inundation periods, which create permanent anaerobic conditions, inhibit nitrification and thus severely curtail nitrogen removal.

The efficiencies of nitrogen and phosphorus removals indicated in Table 3.3 are based on values obtained over the last 15-20 years in comparable experimental systems under a variety of environmental conditions. The percentage nitrogen removals appear to be in very good agreement with agricultural fertilization experiences, where under optimum conditions 2-30% of applied fertilizer nitrogen is lost to drain water.[10,72] The estimated relatively low percentages of phosphorus removals in the overland runoff and rapid infiltration ponds are due to possible losses of phosphorus in surface runoff and in water percolating downward under the two different systems of wastewater application, respectively. The low efficiency of phosphorus removal in the overland runoff system is attributed to low retention time of wastewater in the clay-clay loam soils, which results from low water infiltration capacities, and to increasing chances of soil erosion, which carries large amounts of phosphorus associated with soil particles in surface runoff. It is estimated that protection of surface soil with a continuous vegetation or litter cover very sharply decreases surface erosion. However, it should be kept in mind that the excessive eutrophication of the surface water is attributed to man-induced nutrients originating to a large extent from land runoff associated with agricultural and private gardening fertilization practices.

The indicated high phosphorus per cent leached to groundwater in the rapid infiltration ponds is based on the low fixation capacity of sand. However, in sands containing iron or aluminum oxides surface coatings or calcium carbonates, the efficiency of sand for phosphorus removal could be higher than 95%. The spray irrigation system for land disposal of wastewater in silt loam soils with moderate or relatively high water infiltration capacity under a continuous vegetation cover is by far the best system for nitrogen and phosphorus removals.[2,6,73] However, even this system would fail to give the expected results if a judicious approach were not constantly maintained. Such an approach must include: (1) continuous efforts to maintain soil conditions conducive to optimum nitrification-denitrification rates for nitrogen removal as N_2 or NO_X, (2) maintaining a permanent vegetation or litter cover and avoidance of irrigation during heavy rainfalls to minimize losses of nitrogen and phosphorus through surface soil erosion, (3) abstaining from wastewater irrigation under frozen soil conditions, and (4) maintaining a highly productive soil, which is ultimately the best guarantee for optimum conditions for land disposal of wastewater. The maintenance of a highly productive soil, which at the same time should be conducive to high nitrogen losses through denitrification, is not an easy job and requires continuous cooperation between soil and water science specialists. Soil amendments such as lime and organic matter applications should be considered continuously to maintain optimum pH and carbon/nitrogen ratios for maximum removals of phosphorus and nitrogen.

REMOVAL OF ORGANIC TOXICANTS

Organic toxicants are those organic impurities that resist the forces of biological, chemical and physical degradation, thus remaining in soils and ground and surface water for considerable periods of time. These, to name a few, include the chlorinated hydrocarbon pesticides, chlorobiphenyls, phthalates and phenolic compounds. All of these and others have been detected in sewage effluents in various concentrations and proportions.[73,74] The amount of the organic toxicants that will reach the land disposal sites cannot be precisely estimated because of the lack of analytical data.[75] However, it is safe to assume that their concentrations in the effluent are variable and not much smaller than 1 mg/l. At this level, the soil filter is quite efficient in preventing removal to ground and/or surface water. However, losses through surface runoff should not be underestimated.

The fate of organic toxicants in soils is determined by a large number of processes, including adsorption by the soil colloids, plant uptake, microbial degradation, chemical breakdown, volatilization and leaching. In a soil, all of these processes may be operating simultaneously.[76,77] The adsorption of trace organics by soil colloids (the most important renovation mechanism) is directly influenced by such factors as the colloid type, the physiochemical nature of the toxicant, the reaction of the medium, the nature of the saturating cation on the colloid exchange site, and temperature. Of all variables involved in the adsorption reaction of organics in soils, organic matter has been recognized as the most important.[77] The spectrum of organic toxicants is so large that no easy generalizations on microbial degradation or reactivation in soils can be made.

REMOVAL OF TRACE METALS

Four soil components–clay minerals, hydrous oxides of iron, aluminum and manganese, organic matter and the biota–are readily identified as the primary factors controlling soil solution concentrations of trace metals in general. Zinc, copper, nickel and cadmium are given special consideration since these metals are potentially toxic to animals and plants. Normally sludge removes most of the trace elements from solution, and an effluent from a domestic sewage contains very small concentrations, (<100 μg/l) of the most toxic metals such as cadmium.[73] Suggested mechanisms of control of the soil solution concentration of heavy metals include (1) adsorption by the layer silicates through surface sorption, surface complex ion formation, lattice penetration, and ion exchange;[78] (2) metal fixation and adsorption by organic matter; (3) surface sorption or surface precipitation on carbonates and/or silicates; (4) precipitation as the discrete oxide or hydroxide; and (5) biological uptake and immobilization.[79,80] With microconcentrations of heavy metals sorption reactions may predominate, while precipitation frequently occurs with macroconcentrations. Several of the sorption mechanisms of heavy metals are proposed to account for the fixation of metals in the sense that they cannot be extracted with salt solutions.[81,82]

The insoluble hydrous oxides of manganese and iron, alone or in combination with organic matter, which are nearly ubiquitous in soils provide one of the principal controls of heavy metals (micronutrients), such as nickel, copper, cobalt and zinc, in soils. The common occurrence of these oxides as coatings allows the oxides to extend chemical activity far out of proportion to their total concentrations. The principal factors that control the sorption and desorption of the hydrous oxide–occluded heavy metals are Eh, pH, concentration of the metal of interest, concentration of competing metals, concentration of other ions capable of forming inorganic complexes and organic chelates. Of these factors, pH and Eh are probably the most significant because they are responsible for the retention of the applied metals in the top few centimeters of soils.[42,83] Thus surface runoff from land disposal sites represents a potential source of trace metals to water bodies. Surface enrichment of trace metals, attributable in part to man's activities, exists in modern surficial sediments of many lakes,[84] and part of this sediment trace metal component has undoubtedly derived from flow of water over soils. The potential of trace metal leached down the soil profile should be minimal when domestic sewage effluents are applied to the land, particularly on soils rich in organic material.[80,85]

REMOVAL OF MAJOR CATIONS AND ANIONS AND TRACE ELEMENTS

The concentration of cations such as sodium, potassium, calcium, magnesium and ammonium in soil solution is controlled primarily by reversible ion exchange reactions, adsorption, plant uptake and for calcium by the formation of insoluble solid phases such as $CaCO_3$ and $CaSO_4$. Charge and size of cations are the main factors governing the relationship between one cation and another in the soil solid-phase solution system. As a consequence of the negative charge of the soil, the solution in close proximity to the clay plates differs from the composition of the soil solution outside the sphere of influence of the negative electric field.[40] Monovalent cations are bound less than divalent cations and thus swarm out farther away from the clay plates than divalent cations. Although the interaction of cations in soils is mainly governed by charge effects, specific forces are always present. Thus the strong and sometimes essentially irreversible binding between cations and the solid phase may be due to a high binding energy of the bond or to mechanical factors preventing the movement of ions once they are bound. Potassium and ammonium "fixation" by illite clay minerals is probably the result of both electrostatic binding and mechanical hindrance.

Sopper[51,52] has shown that renovation of a waste effluent can be achieved for many cations in a 4-foot (1.2-m) soil profile. Sustained capacities for the soil to renovate potassium, calcium and magnesium were 83, 59, and 53%, respectively. Although potassium and copper losses due to leaching in soils are small compared to the total quantity of these elements in the topsoil, these losses may be appreciable when compared to other elements. The losses of cations are considerably reduced when the land is cropped.

The quantitative description of the solid-phase soil-solution relationship of anions requires individual treatment. Those anions bound by the soil generally reflect both a sharing of electrons and covalent bonding in contrast to the electrovalent bonding of the cations. The concentration of anions in soil solution is affected by the nature of the solid phase, and the overall negative charge of the adsorption complex causes anions to be repulsed or "negatively adsorbed" in close proximity to the charged sites. Anion binding is the most important phenomenon determining the soil solution concentration of anions. Different anion binding mechanisms that are affected by pH and salt concentration operate in the soil. Anions such as NO_3, Cl and SO_4 do not normally react with the soil solid phase; however, retention of SO_4 may be produced in acid soils high in iron and aluminum oxides.[86,87]

SUMMARY OF RENOVATION MECHANISMS

To demonstrate the interaction of parameters, effluent components, soil renovation mechanisms, and land application methodologies, the matrix in Table 3.4 was developed. The information included in this matrix was derived from the previous sections on the removal of individual effluent components. The relative scale 0-4 is intended to show the relative importance of a number of soil-plant processes in land disposal renovation of individual effluent components. Where adequate data do not exist, the values shown represent the best judgment of the authors. Some soil processes were put under only one of seven groups of effluent renovation mechanisms whereas

Table 3.4. Land Disposal Renovation Mechanisms and Their Relative Predominance in Spray Irrigation (SI), Overland Runoff (OR) and Rapid Infiltration (RI) Systems[a]

Effluent Constituent	Plant Up-Take[b]			Microbial Degradation and/or Immobilization[c]			Gaseous Losses[d]			Adsorption Precipitation			Ion Exchange			Mechanical Retention		
	SI	OR	RI	SI	OR	RI	SI	OR	RI	SI	OR	RI	SI	OR	RI	SI	OR	RI
BOD	0	0	0	4	4	2	0	0	0	2	2	2	0	0	0	4	3	4
Suspended solids	0	0	0	4	4	2	0	0	0	2	2	2	0	0	0	4	3	4
Nitrogen	4	3	0	2	2	1	4	4	4	1	1	0	2	1	1	1	0	1
Phosphorus	4	3	0	1	1	0	0	0	0	4	4	4	1	1	1	1	0	1
Trace metals	1	1	0	2	2	1	0	0	0	4	4	4	1	1	1	0	0	0
Organic toxicants	0	0	0	3	2	1	1	1	1	4	4	4	0	0	0	0	0	0
Bacteria and viruses	0	0	0	3	4	3	0	0	0	2	2	2	0	0	0	4	3	4
Major cations	4	3	1	0	0	0	0	0	0	4	4	4	2	1	2	0	0	0
Major anions	2	3	0	0	0	0	0	0	0	4	4	4	0	0	0	0	0	0

[a]A 0-4 scale is used to express the relative importance of the proposed mechanisms and is derived from ranking the processes controlling the soil solution composition of a given component: "0" indicates that the process is inoperative or relatively insignificant whereas "4" designates the major mechanism.

[b]Removal by harvest crops.

[c]Includes decomposition of organic material and loss of CO_2 to atmosphere and kill or other inactivation of pathogenic organisms.

[d]Loss to atmosphere of: (1) N_2 and N oxides formed by biological and/or chemical denitrification, (2) volatile organic compounds, and (3) NH_3.

in reality they could just as well be a part of another group. Microbial degradation serves as an example: CO_2 gas is lost, thus allowing the mechanism to be considered under "gaseous losses." When such arbitrary separations of soil processes are employed, the intent is to emphasize the importance of a single process that otherwise would be overlooked if it is grouped with others of a similar function.

POTENTIAL EFFECTIVENESS OF LAND DISPOSAL

The extent of renovation of a chlorinated (domestic-industrial) secondary effluent of model composition (Table 3.5) is estimated for model soil types and a climate similar to that prevailing in the southern Great Lakes. The

Table 3.5. Assumed Water Quality of Secondary Effluent and Drinking-Irrigation Water Standards[a]

	Effluent[b]	Drinking Water	Irrigation Water
BOD	25	–	–
COD	70	–	–
Organic toxicants[c]	<1	1	–
Total N	20	–	–
NH_4-N	12	0.5	–
NO_3-N	5	10	–
Total P	10	–	–
Na	50	–	–
K	14	–	–
Ca	24	–	–
Mg	17	–	–
Cd	0.1	0.01	0.01
Cr	0.2	0.05	–
Cu	0.2	1.0	0.2
Fe	0.5	0.3	–
Pb	0.1	0.05	5
Mn	0.5	0.05	2
Ni	0.2	–	0.2
Zn	0.3	5	2
B	1	–	0.75
Suspended solids	25	5	–
pH	7 ± 0.5	6 - 8.5	–

[a]Values in mg/l except for pH.

[b]Microorganisms are assumed to be typical of chlorinated municipal secondary effluents.

[c]Organic toxicants are assumed to be included in the carbon chloroform extract.

model conditions assumed for the purpose of this discussion are summarized under the three basic approaches to land application (spray irrigation, overland runoff and rapid infiltration) in Table 3.6.[2,6,9,73,88,89]

A matrix depicting the expected renovation effectivenesses of the wastewater disposal methodologies is constructed in Table 3.7 on the basis of past experiences with similar systems described in the previous sections. The matrix takes into account the site conditions, effluent amounts, application rates and other land disposal characteristics described in Table 3.6 and the mechanisms of renovation (Table 3.4) for individual effluent components at the concentratives given in Table 3.5.

The removal of BOD and SS in spray irrigation and rapid infiltration is expected to be very effective–99%–largely due to the high filtering effectiveness of the soil mantle and maintenance of aerobic conditions by alternating wet and dry periods. These removal efficiencies have been attained with loading rates thousands of times greater than proposed here, so nearly complete removal is expected. Overland runoff is expected to provide a higher residual in the seepage water leaving the field because some of the filtering capacity of the soil is lost with horizontal transport of the wastewater. Since suspended solids are largely organic in secondary wastewater, their behavior should about parallel that of BOD.

The percentage of nitrogen removals (60-90%) for the spray irrigation and overland runoff is in good agreement with experiences in agricultural fertilization. The low efficiency of phosphorus removal in the overland runoff system is attributed to short retention time of wastewater in soils. The indicated higher percentage of phosphorus leached to groundwater in the rapid infiltration ponds is based on the lower fixation capacity of coarse-textured soils.

Efficiencies of nitrogen removal may be increased to more than 90% and maintained for many years by additions of organic carbon. A high carbon/nitrogen ratio accumulates nitrogen in the soil which minimizes loss through leaching. Increased soil organic matter also maintains or improves soil structure, increases the potential for plant uptake, and aids aeration and water flow processes.

Heavy metals are efficiently renovated by spray irrigation as 100% of the wastewater infiltrates the soil column, thus allowing both time and surface area for ion exchange and adsorption reactions. Overland runoff does not permit sufficient contact for heavy metals to be adsorbed, exchanged, or reacted chemically in the soil to obtain the degree of renovation expected with rapid infiltration or spray irrigation. Assuming soils have a few colloids and the system is relatively aerobic, metals are renovated approximately 50-90% by rapid infiltration ponds. Organic toxicants are expected to be held in soil and not to penetrate into the water table in any appreciable

Table 3.6. Assumed Model Land Disposal Conditions and Characteristics for Treating Chlorinated Industrial-Domestic Secondary Effluent[a]

Item	Spray Irrigation	Overland Runoff	Rapid Infiltration
Soil type	medium texture soils, silt loam	clay, clay loam	loamy sand to sandy loam
Permeability of the most impermeable subsoil horizon to 60 in. (1.5 m)	moderately rapid to moderate; >0.6 in. (1.5 cm)/hr	very low; <0.2 in. (0.5 cm)/hr	rapid
Infiltration	moderately rapid to moderate	slow-very slow; <0.2 in. (0.5 cm)/hr	rapid to very rapid; ≥1 in. (5 cm)/hr
Soil drainage	moderately well drained	poorly drained	excessively drained
Soil depth	uniformly >5 feet (1.5 m)	>2 feet (0.6 m)	>10 feet (3 m)
Effective travel distance	preferred 10 ft (3 m) or more	surface travel of 150 feet (46 m)	subsurface > 200 ft (60 m)
Slope	up to 8 ft (2.4 m)	5 ft (1.5 m)	——
Waterholding capacity to 60 in. (1.5 m)[b]	>6 in. (15 cm)	——	——
Depth to ground water	> 5 ft (1.5 m) to seasonal high groundwater table	> 3 ft (0.9 m)	15-20 ft (4.5-6 m) or more
Vegetation	grass, year-round	permanent growing grass	vegetated
Climate	similar to that of Great Lakes	Great Lakes	Great Lakes
Application amount and rate	2 in. (5 cm)/wk once a week @ 0.2 in. (0.5 cm) to 0.25 in. (0.64 cm)/hr	2 in. (5 cm)/wk 4 times a week at 0.5 in. (1.3 cm)/day	60 in. (150 cm)/wk
Land required for 1-mgd (378,500 m^3/day) disposal	129 acres (52 ha)	129 acres (52 ha)	4.3 acres (1.7 ha)
Resting period between applications	3-6 days	3 days or more	14 days wet - 14 days dry
Duration	9 mo. Mar-Nov.	9 mo. Mar-Nov	year-round
Application technique	spray	spray	surface

[a]The wastewater composition indicated in Table 3.5 was taken into account in choosing the effluent application amounts shown.

[b]Water-holding capacity in inches from the soil column is the depth of the layer of water that would be formed if all water in the soil that can be used by plants were concentrated at the soil surface.

Table 3.7. Estimated Percentage Renovation Effectiveness of Land Disposal Techniques[a]

	Spray Irrigation[b]	Overland Runoff[b]	Rapid Infiltration Ponds
BOD	99	80	99
SS	99+	80	99
N	70-90	60-90	30-80
P	95-99	60-80	50-90
Trace metals	95-99	60-80	50-90
Organic toxicants	99	60-80	90
Viruses and bacteria	99+	90	99+
Cations	50-75	30-50	30-75
Anions	0-50	0-10	0-50

[a]Modified from Reference 73. Renovation effectiveness is expressed as percentage of the original effluent constituent content that is retained in the soil, utilized by vegetation or volatilized under the conditions outlined in Tables 3.5 and 3.6.

[b]Removal by harvest crop is included.

amount. Overland runoff is considered least effective in organic toxicants removal. The rapid infiltration technique seems to be nearly as effective as spray irrigation for organic toxicant removal.

Bacteria and viruses are removed from wastewater systems by the upper soil mantle (1 cm-2 m depending on soil type) very efficiently when applied by spray irrigation and disposed of by rapid infiltration pond systems. However, when processed by the overland runoff method, viruses and bacteria are probably removed somewhat less efficiently (90% *vs.* 99%). Although the wastewater in an overland runoff system flows a few hundred feet through and over the top few inches of soil and through the litter and vegetative cover, there is a greater possibility for the bacteria and viruses to be carried to receiving waters by the overland flow method than by methods utilizing vertical percolation of the wastewater through the soil.

Total dissolved solids as ions are most effectively renovated by spray irrigation or rapid infiltration ponds. The residence time in association with an adequate surface of colloidal soil particles should be sufficient to achieve the degree of renovation suggested. Overland runoff is less efficient in removing total dissolved solids because only the soil surface and impeding plant residues are effective in the chemical reaction.

POTENTIAL ENVIRONMENTAL IMPACTS OF LAND DISPOSAL

In estimating potential environmental impacts from treatment of secondary effluent on land by spray irrigation, overland runoff and rapid infiltration ponds, a region with a climate similar to that of the southern Great Lakes is to be considered. The lower Great Lakes area of southern Michigan is composed mostly of glacial till and subsurface glacial moraines. Application of 80 in. (203 cm) to 240 feet (78 m) per year of effluent superimposed on 30 in. (76 cm) or so of natural rainfall may have a significant environmental effect. The impact of leachate or runoff water containing residual pollutants on adjacent water supplies and aquatic ecosystems must be considered.

Eutrophication

A potential impact of land disposal of secondary wastewater on the surrounding environment is accelerated eutrophication of contiguous surface waters. Accelerated eutrophication can occur in either standing or running water, but is more common in standing water. However, in both environments planktonic or attached plants assimilate available nutrients according to their requirements, which are in part dictated by their finite capacity for uptake and other growing conditions such as light and temperature. If growth and population increase in either standing or running water is limited (or restricted) by the quantity of an available nutrient, then addition of such a nutrient will stimulate growth and result in a larger mass of algae, possibly reaching nuisance proportions if the nutrient increase is great enough. Large masses of nuisance plants indicate accelerated eutrophication.

The two nutrients most often cited as causes of accelerated eutrophication are nitrogen and phosphorus. Planktonic algal growth rate can be limited by many nutrients, including carbon, iron, and several other trace elements. However, a principal reason why nitrogen and phosphorus, and particularly phosphorus, have been most often linked to accelerated eutrophication is that, relative to needs for algal biomass increase, these elements are the most scarce naturally. Phosphorus probably has more "sinks" and fewer sources than nitrogen, so is most often considered the nutrient most responsible for long-term increase in algal mass and productivity, which are characteristic of eutrophication.

Significantly increased loading of nitrogen and/or phosphorus would be most apt to result in nuisance levels of algae if added to standing water than if added to running water. This is probably more true for shallow than deep lakes. Even though growth is possible in the lighted surface water only, the incoming nutrients are usually efficiently scavenged and converted to algal mass. Because current is very effective in maintaining transport of nutrients to the bottom-attached algae in streams, growth is often maximized at lower concentrations than in lakes. Thus, further increase in nutrients in streams has little effect except when existing concentrations are very low or current is very sluggish. However, if travel time is sufficiently long for nutrient content to be eventually depleted by uptake, then plant growth could again be limited downstream. But in the meantime nutrients have been distributed over a long distance, particularly if flow is great. In the case of sluggish flow or standing water, nutrients are detained, allowing ample time for more complete utilization in a smaller area. Thus, larger algal masses are apt to result from increased nutrient concentrations.

The impact of treated wastewater on eutrophication on contiguous surface waters is considered herein from the standpoint of total annual loading and possible concentration increase of nitrogen and phosphorus. Vollenweider has reviewed trophic characteristics of several lakes in the world and, according to their measured nitrogen and phosphorus loading rates and mean depths, has suggested admissible and danger limits for these two nutrients.[90] By significantly exceeding the danger limits in annual loading of nitrogen and/or phosphorus in a lake of known depth, eutrophication will probably be accelerated.

A site of 6,450 acres (2,610 ha) with 78 in. (200 cm) of model wastewater applied per year in the case of spray irrigation and overland runoff and a site of 215 acres (87 ha) receiving 260 ft (80 m) model effluent per year (rapid infiltration ponds) are used as guides to estimate nutrient loading to contiguous water bodies from groundwater recharge or overland runoff (Table 3.8). These land disposal sites are characteristic of a city of 500,000 people. The mean concentrations of nitrogen and phosphorus in wastewater (Table 3.5) and rainfall (Table 3.9) were used to calculate

Table 3.8. Annual Total Effluent Application for a City of 500,000 People[a]

Disposal System	Required Area		Application Period, Weeks	Total Effluent Applied	
	Acres	Hectares		Acre-Feet	10^3 m^3
Spray irrigation	6,450	2,610	39	41,890	51,671
Overland runoff	6,450	2,610	39	41,890	51,671
Rapid infiltration	215	87	52	55,860	68,903

[a]Assume a 100 gal (0.3785 m^3)/day/person wastewater production. Application rates are from Table 3.6.

Table 3.9. Constituent Concentrations in Soils, Surface Waters, and Precipitation

Constituent	Soils ppm[a]	U.S. River Basins[b] All U.S. μg/l	U.S. River Basins[b] Alaska μg/l	U.S. Precipitation[c] μg/l
Zn	10-300	64	28	107
Cu	2-100	15	9	21
Pb	2-200	23	12	34
Cd	0.01-0.7	< 1	–	<5
Organic toxicants[d]	–	<100	–	<200
N[e]	10^3-2×10^4	1,600	–	500
P[e]	10^2-1.3×10^3	200	–	25

[a]From Swaine.[91]
[b]Trace metal data from Kopp and Kroner.[92] The Cd value is from Durum *et al.*[93]
[c]From Lazrus *et al.*[94] The Cd value was computed as 5% of the Zn value.
[d]Values are arbitrary. They were calculated as 1-2% of COD.
[e]Nitrogen and phosphorus values for U.S. rivers and precipitation are within the limits reported in literature.

total nitrogen and phosphorus annual loads to the disposal sites. Potential evapotranspiration was assumed to be equal to rainfall. Table 3.10 provides estimates of the total amounts of nitrogen and phosphorus that are lost to groundwater and/or surface runoff from the three application techniques under average and maximum renovation effectivenesses. Under average renovation conditions phosphorus loading from overland runoff is much greater than that from spray irrigation. Phosphorus and nitrogen loading from the rapid infiltration pond system are of dramatic proportions. From Table 3.10 it is also clear that loading to standing water from land disposal may produce a significant effect with phosphorus because in two techniques of disposal the ratio of nitrogen/phosphorus by weight is less than that required by algae, 7/1; relatively more phosphorus would be available than nitrogen. This is particularly important because natural waters generally have a nitrogen/phosphorus ratio much higher than 7/1. With spray irrigation the opposite is true due to the high efficiency of phosphorus removal by that technique.

Considering Vollenweider's[90] suggested danger limits and a hypothetical contiguous ecosystem of standing water with a mean depth of 65 feet (19.8 m), Table 3.11 gives estimates of the area in 1000's of acres that would be required to dissipate the nutrients without causing accelerated eutrophication. The danger limits in lakes are 37 lb nitrogen/ac yr and 2.7 lb phosphorus/ac yr. The lower estimate of treatment efficiency by rapid infiltration results in the largest estimate of area needed for nutrient dissipation–about 280,000 acres (113,000 ha) if the groundwater drainage from a 215 acre (87 ha) disposal site reached standing surface water averaging 65 feet (19.8 m) deep. Even with spray irrigation, nearly 20,000 acres are required. Of course loss of nutrients from the site to surface water through groundwater or seepage from overland runoff is not the only path. Erosion from the site surface, which is typical in agricultural areas, could be another important path for nutrient loss to surface water. Even though treatment efficiencies for nitrogen and phosphorus removal are relatively high by most techniques, sizable loads may still escape, and if the ground or surface water runoff reaches nearby lakes, accelerated eutrophication may occur if the lakes are not large or deep enough for adequate dissipation of added nutrients. In Table 3.12 the estimated losses of nitrogen and phosphorus from spray irrigation and overland runoff sites are compared to those from harvested crop areas in the U.S.[95] Characteristically all the values except those for phosphorus loss from overland runoff are comparable.

From the concentrations of nitrogen and phosphorus in the renovated water (Table 3.10) it becomes apparent that the addition of significant quantities of reclaimed water to receiving streams in most areas would no doubt raise concentrations of nitrogen and phosphorus, with the possible exception of phosphorus in spray irrigation. In streams with adequate light penetration and relatively low minimum concentrations, reclaimed water may result in nuisance levels of algae.

Potential Toxicity for Trace Metals and Organic Toxicants

As indicated by the calculated concentration of trace metals and organics in the renovated water (Table 3.10) the potential of groundwater contamination is real when U.S. public health standards are considered. The cadmium concentrations in the renovated water from overland runoff and rapid infiltration exceeded the public water standards (0.01 mg/l) probably as a result of the industrial component of the applied effluent. The 100 μg Cd/l assumed to be present in the effluent was probably much in excess of expected and measured concentrations.[84]

Nitrate in Domestic Water

The USPHS in 1962 Drinking Water Standards recommended a limit of 45 mg/l nitrate as NO_3 (10.17 mg/l NO_3-N). The recommended limit for

Table 3.10. Annual Loads, Leachates, and Renovated Water Concentrations
Constituent annual loads, amounts leached and/or eroded and resultant concentrations in ground and/or surface renovated water under maximum and average renovation effectivenesses (RE) for 50-mgd (189,250 m^3/day) spray irrigation, overland runoff, and rapid infiltration model systems serving a city of 500,000 people[a]

	Annual Load[b]		Leached or Eroded Annually[c]				Renovated Water	
			Total lb		lb/acre		mg/l[d]	
Constituents	Total 1,000 lb	lb/acre	Max. RE	Avg. RE	Max. RE	Avg. RE	Max. RE	Avg. RE
Spray Irrigation								
Nitrogen	2,300	356	230,000 (90)[e]	460,000 (80)[e]	35.7	71.3	2.0	4.0
Phosphorus	1,140	176	11,000 (99)	34,200 (97)	1.8	5.3	0.1	0.3
Organic toxicants	123	19.4	1,230 (99)	–	0.2	–	0.01	–
Zinc	39	5.9	390 (99)	1,170 (97)	0.1	0.2	0.003	0.010
Copper	24	3.6	240 (99)	720 (97)	0.04	0.1	0.002	0.006
Lead	13	2.0	130 (99)	390 (97)	0.02	0.06	0.001	0.003
Cadmium	12	1.8	120 (99)	360 (97)	0.02	0.06	0.001	0.003
Overland Runoff								
Nitrogen	2,300	356	230,000 (90)	575,000 (75)	35.7	89.1	2.0	5.1
Phosphorus	1,140	176	228,000 (80)	342,000 (70)	35.3	53.0	2.0	3.0
Organic toxicants	123	19.4	24,600 (80)	36,900 (70)	3.8	5.7	0.22	0.32
Zinc	39	5.9	7,800 (80)	11,700 (70)	1.2	1.8	0.068	1.03
Copper	24	3.6	4,800 (80)	7,200 (70)	0.74	1.1	0.042	0.063
Lead	13	2.0	2,600 (80)	3,900 (70)	0.40	0.61	0.023	0.034
Cadmium	12	1.8	2,400 (80)	3,600 (70)	0.37	0.56	0.021[f]	0.032[f]

Rapid Infiltration								
Nitrogen	3,039	14,134	607,800 (80)	1,367,550 (55)	2,827	6,361	4.0	9.0
Phosphorus	1,519	7,065	151,900 (90)	455,700 (70)	707	2,120	1.0	3.0
Organic toxicants	152.2	708	15,220 (90)	–	71	–	0.1	–
Zinc	45.9	214	4,590 (90)	13,770 (70)	21	64	0.030	0.091
Copper	30.4	142	3,040 (90)	9,120 (70)	14	42	0.020	0.060
Lead	15.2	71	1,520 (90)	4,560 (70)	7	21	0.010[f]	0.030[f]
Cadmium	15.2	71	1,520 (90)	4,560 (70)	7	21	0.010[f]	0.030[f]

[a]Data was compiled from Tables 3.5, 3.7, 3.8 and 3.9. To calculate the annual constituent load, the contribution of rainfall (from Table 3.8), assumed to be 30 in./yr (76.2 cm/yr), was added to the effluent load.

[b]To convert lb to kg and lb/acre to kg/ha, multiply by 0.4536 and 1.12, respectively.

[c]The percentage maximum and average renovation effectivenesses are based on the values in Table 3.7.

[d]The constituent concentrations are based on the total volume of the applied effluent. Potential evapotranspiration was assumed to equal precipitation.

[e]Values in parenthesis are estimated per cent renovation from Table 3.7.

[f]Constituent concentrations indicate that drinking water standards were exceeded (Table 3.5).

Table 3.11. Size of Lakes to Dissipate Nitrogen and Phosphorus Loads. Estimated 1,000's of acres of standing surface water averaging 65 feet (19.8 m) in depth that may be necessary to dissipate the annual nitrogen and phosphorus loadings from model land renovated effluents without exceeding the tentative danger loading rates of nitrogen (37 lb/ac yr or 41 kg/ha yr) and phosphorus (2.7 lb/ac yr or 3.0 kg/ha yr) for eutrophication as proposed by Vollenweider.[90]

	Required 1,000 Acres of Surface Water Under[a]					
Constituent	Maximum Renovation Effectiveness	(%)	Minimum Renovation Effectiveness	(%)	Average Renovation Effectiveness	(%)
			Spray Irrigation			
Nitrogen	6.2	(90)	18.6	(70)	12.4	(80)
Phosphorus	4.2	(99)	21.1	(95)	12.7	(97)
			Overland Runoff			
Nitrogen	6.2	(90)	24.8	(60)	15.5	(75)
Phosphorus	84.4	(80)	169.0	(60)	126.7	(70)
			Rapid Infiltration			
Nitrogen	16.4	(80)	57.6	(30)	37.0	(55)
Phosphorus	56.3	(90)	281.3	(50)	168.8	(70)

[a]The annual nitrogen and phosphorus loads from the Model Land Disposal Systems were calculated from the annually applied nitrogen and phosphorus (Table 3.10) after correcting for the indicated renovated effectivenesses (Table 3.7). To convert acres to hectares, multiply by 0.4047.

European drinking water was set in 1961 by the World Health Organization at 50 mg/l NO_3 (11.3 mg/l NO_3-N). These limits were adopted because of the relationship between high nitrates in drinking water and nitrate poisoning in infants (infant methemoglobinemia). Low removal efficiencies of nitrates from wastewater in a land disposal system could result in contamination of the groundwater and present a potential health hazard if the water is reclaimed for unrestricted municipal use. If all nitrogen in the model wastewater (Table 3.5) were converted to nitrate, the concentration of NO_3-N (20 mg/l) would be twice the recommended limit for drinking water. However, maintenance of nitrogen removal efficiencies commensurate with well-managed land disposal systems ($\geq$80%) coupled with dilution from rainfall and groundwater, would insure against a health hazard.

Table 3.12. Nitrogen and Phosphorus Losses.
Harvested areas vs. model land disposal systems. Losses of nitrogen and phosphorus from harvested crop areas in the U.S. as compared to those obtained from model land disposal systems of spray irrigation and overland runoff[a]

	Loss, lb/ac yr[b]	
	Nitrogen	Phosphorus
Intertilled crops		
Leaching	17.1	–
Erosion	48.1	21.0
Annual crops not intertilled		
Leaching	32.5	–
Erosion	11.1	4.9
Biennial and perennial crops		
Leaching	23.0	–
Erosion	24.2	10.6
Spray irrigation[c]		
90% renovation	35.7	–
80% renovation	71.3	–
99% renovation	–	1.8
97% renovation	–	5.3
Overland runoff[c]		
90% renovation	35.7	–
75% renovation	89.1	–
80% renovation	–	35.3
70% renovation	–	53.0

[a]From Lipman and Conybeare.[95]

[b]To convert lb/ac yr to kg/ha yr, multiply by 1.12.

[c]From Table 3.10.

In poorly managed situations removal efficiencies could be as low as 50%, which would result in NO_3-N content exceeding the 10 mg/l limit.

CONCLUSIONS

Wastewater application to soil-plant systems can provide an efficient treatment only under very strict conditions of control of both the quality of land-disposed effluent and land management. Excessive leaching of NO_3-N to groundwater, escape of toxic substances, and runoff of phosphorus and nitrogen appear to be the three potential problem areas. The spray irrigation system for land disposal of wastewater in silt loam soils with moderate to relatively high water infiltration capacity under a continuous vegetation cover is by far the best system for nitrogen, phosphorus and trace toxicants removal. However, even this system would fail to give the expected results if a judicious approach were not constantly maintained. Such an approach would include: (1) continuous effort for maintaining soil conditions conducive to optimum nitrification-denitrification rates for nitrogen removal as N_2 or NO_X, (2) permanent vegetation cover and avoidance of irrigation during heavy rainfall to minimize leaching of NO_3 and losses of nitrogen and phosphorus through surface soil erosion, and (3) maintenance of a highly productive soil, which is ultimately the best guarantee for optimum conditions for land disposal of wastewater.

REFERENCES

1. Westman, W. E. "Some Basic Issues in Water Pollution Control Legislation," *Amer. Scientist* **60**, 767 (1972).
2. Sopper, W. E. and L. T. Kardos, Eds. *Recycling Treated Municipal Wastewater and Sludge Through Forest and Cropland*, Proceedings of a Symposium, Pennsylvania State University, August, 1972 (University Park, Pa.: Pennsylvania State University Press, 1973).
3. Alexander, M. *Microbial Ecology.* (New York: John Wiley and Sons, Inc., 1971).
4. Alexander, M. *Introduction to Soil Microbiology.* (New York: John Wiley and Sons, Inc., 1961).
5. Willrich, R. L. and G. E. Smith, Eds. *Agricultural Practices and Water Quality.* (Ames, Iowa: The Iowa State University Press, 1970).
6. National Assoc. of State Universities and Land-Grant Colleges. *Recycling Municipal Sludges and Effluents on Land.* Proceedings of a Joint Conference, July 9-13, 1973, Champaign, Ill.
7. Bauer, W. J. *Heavy Metals in Soils and Crops,* Vol. 4. (Chicago, Ill.: Soil Enrichment Materials Corporation, 1972).
8. Brady, N. C., Ed. *Agriculture and the Quality of Our Environment.* AAAS Publication 85, Washington, D.C. (1967).
9. Pound, C. E. and R. W. Crites. *Wastewater Treatment and Reuse by Land Application*, Vol. 1. Summary EPA-660/2-73-006a, Washington, D.C. (1973).
10. Fried, M. and H. Broeshart. *The Soil-Plant System.* (New York: Academic Press, 1967).
11. Wilde, S. A. *Forest Soils.* (New York: The Ronald Press Co., 1958).
12. Miller, R. H. "Soil Microbiological Aspects of Recycling Sewage Sludges and Waste Effluents on Land," *Proceedings of a Joint Conference on Recycling Municipal Sludges and Effluents on Land.* Champaign, Ill. July 9-13, 1973, pp. 79-90.
13. Miller, R. H. "The Soil as a Biological Filter," in *Recycling Treated Municipal Wastewater and Sludge Through Forest and Cropland,* W. E. Sopper and L. T. Kardos, Eds. (University Park, Pa.: Pennsylvania State University Press, 1973).

14. Drewry, W. A. and R. Eliassen. "Virus Movement in Ground Water," *J. Water Poll. Control Fed.,* **Pt. 2**, R257 (1968).
15. McGauhey, P. H. "Manmade Contamination Hazards," *Ground Water* **6**(3), 10 (1968).
16. Ewing, B. B. and R. I. Dick. "Disposal of Sludge on Land," In *Water Quality Improvement by Physical and Chemical Processes,* E. F. Gloyna and W. W. Eckenfelder, Jr., Eds. (Austin, Tex.: University of Texas Press, 1970), pp. 394-408.
17. Marculeseu, I. and Drucan. *Investigations Using Labelled Bacteria in the Study of Irrigation with Sewage.* (Bucharest: Stud. Prot. Epur. Apel, 1962), pp. 59-66.
18. Law, J. P., Jr., R. E. Thomas and L. H. Myers. "Nutrient Removal from Cannery Wastes by Spray Irrigation of Grassland," U.S. Dept. Interior, FWPCA, 16080-11/69 (1969).
19. Merrell, J. C., Jr., A. Katko and H. E. Pantler. "The Santee Recreation Project, Santee, Calif.," Summary Report, 1962-64, PHS Publ. No. 999-WP-27 (1965).
20. Bouwer, H. "Groundwater Recharge Design for Renovating Waste-Water," *J. SED-ASCE*, **96**, SAI, **59** (1970).
21. Romero, J. C. "The Movement of Bacteria and Viruses Through Porous Media," *Ground Water* **8**(2), 37 (1970).
22. McGauhey, P. H. and R. G. Krone. "Report of Investigation of Travel of Pollution," *Calif. State Water Poll. Bd.* **Publ. 11,** 218 (1954).
23. Rudolfs, W. L., L. Falk and R. A. Ragotzkie. "Literature Review of the Occurrence and Survival of Enteric, Pathogenic and Relative Organisms in Soil, Water, Sewage, and Sludges, and on Vegetation," *Sewage Ind. Waste* **22**(10), 1261 (1950).
24. McGauhey, P. H. and R. G. Krone. "Report of Investigation of Travel of Pollution," *Calif. State Water Poll. Bd.,* **Publ. 11**, 218 (1954).
25. Chambers, C. W. "Chlorination for Control of Bacteria and Viruses in Treatment Plant Effluents," *J. Water Pollution Control Fed.,* **43**(2), 228 (1971).
26. Mallman, W. L. and W. N. Mack. "Biological Contamination of Ground Water," *Ground Water Contamination,* USPHS Tech. Rept. W61-5, 35 (1961).
27. Sepp, E. *Survey of Sewage Disposal by Hillside Sprays*, Bureau of Sanit. Eng., Calif. State Dept. of Pub. Health (1965), 32 pp.
28. Tamm, C. O. and H. G. Ostlund. "Radiocarbon Dating of Soil Humus," *Nature* **185**(4714), 706 (1960).
29. Painter, H. A. "Composition of Sewage and Sewage Effluents," *J. Inst. Sew. Purif.,* **Part 4**, 302 (1961).
30. Bunch, R. L., E. F. Barth and M. B. Ettinger. "Organic Materials in Secondary Effluents," *J. Water Poll. Control Fed.* **33**(2), 122 (1961).
31. Crawford, S. C. "Spray Irrigation of Certain Sulfate Pulp Mill Wastes," *Sewage Indust. Wastes* **30**(10), 1266 (1961).
32. Elazar, D. J. *Greenland-Clean Streams. The Beneficial Use of Waste Water Through Land Treatment* (Philadelphia: Temple Univ. Press, 1971).
33. Jones, J. H. and G. S. Taylor. "Septic Tank Effluent Percolation Through Sands Under Laboratory Conditions," *Soil Sci.* **99**(5), 301 (1965).

34. Laak, R. "Influence of Domestic Wastewater Pretreatment on Soil Clogging," *J. Water Poll. Control Fed.* **42**(8), 1495 (1971).
35. Allison, L. E. "Effect of Microorganisms on Permeability of Soil Under Prolonged Submergence," *Soil Sci.* **63**, 439 (1947).
36. Winneberger, J. H., *et al.* "Biological Aspects of Failure of Septic Tank Percolation Systems–Final Report," Report to FHA, Aug. 31, 1960. Sanit. Eng. Res. Lab., Univ. of Calif., Berkeley, Calif.
37. Mitchell, R., *et al.* "Environmental Conditions Controlling Polysaccharide Production and Degradation by Predominating Microorganisms," Weismann Inst. of Science, Rehovot, Israel (April 1965).
38. Thomas, R. E. and J. P. Law, Jr. "Soil Response to Sewage Effluent Irrigation," Proceedings of the Symposium on the Use of Sewage Effluent for Irrigation, Louisiana Polytechnic Inst., July 30, 1968, 21 pp.
39. Black, C. A. *Soil-Plant Relationships.* (New York: John Wiley & Sons, Inc., 1967)
40. Marshall, C. E. *The Physical Chemistry and Mineralogy of Soils.* Vol. 1 *Soil Materials* (New York: John Wiley & Sons, Inc., 1964).
41. Bailey, G. W. "Role of Soils and Sediment in Water Pollution Control. Part 1," Southeast Water Laboratory, (Washington, D.C.: U.S. Dept. of the Interior, FWPCA, 1968).
42. Stumm, W. and J. J. Morgan. *Aquatic Chemistry*. (New York: Wiley-Interscience, 1970).
43. Bartholomew, W. V. and F. C. Clark, Eds. *Soil Nitrogen.* (Madison, Wisc.: American Soc. of Agronomy, Inc., 1965).
44. Broadbent, F. E. "Factors Affecting Nitrification-Denitrification in Soils," in *Recycling Treated Municipal Wastewater and Sludge Through Forest and Cropland,* W. E. Sopper and L. T. Kardos, Eds. (University Park, Pa.: The Pennsylvania State University Press, 1973).
45. Woldendorp, J. W., K. Dilz and G. J. Kolenbrander. "The Fate of Fertilizer Nitrogen on Permanent Grassland Soils," Proc. 1st General Meeting European Grassland Fed. (1965), 53-76.
46. Cole, D. W., S. P. Gessel and S. F. Dice. "Distribution and Cycling of Nitrogen, Phosphorus, Potassium and Calcium in a Second-Growth Douglas-Fir Ecosystem," in *Primary Productivity and Mineral Cycling in Natural Ecosystems,* AAAS Symp. (Orono, Maine: University of Maine Press, 1968).
47. Gessel, S. P., D. W. Cole and E. C. Steinbrenner. "Nitrogen Balances in Forest Ecosystems of the Pacific Northwest," *Soil Biol. Biochem.* 5, 19 (1973).
48. Tamm, C. O. "Growth and Nutrient Concentrations of Spruces on Various Levels of Nitrogen Fertilization," in *Plant Analysis and Fertilizer Problems,* C. Bould, P. Prevot and J. R. Magness, Eds. (New York: W. F. Humphrey Press, 1964), pp. 344-356.
49. Hallsworth, E. G. and D. V. Crawford, Eds. *Experimental Pedology*. (London: Butterworths, 1965).
50. Reed, S. C. "Wastewater Management by Disposal on the Land," Special Report 171, Cold Regions Research and Engineering Laboratory, Hanover, N.H. (1972).
51. Sopper, W. E. "Disposal of Municipal Wastewater Through Forest Irrigation," *Environ. Poll.* **1**, 263 (1971).

52. Sopper, W. E. "Effects of Trees and Forests in Neutralizing Waste in Trees and Forests and an Urbanizing Environment," A monograph. Univ. of Mass., USDA Printing and Resource Development, Series 17.
53. Sopper, W. E. and L. T. Kardos. "Vegetation Responses to Irrigation with Treated Municipal Wastewater," in *Recycling Treated Municipal Wastewater and Sludge Through Forest and Cropland*, W. E. Sopper and L. T. Kardos, Eds. (University Park, Pa.: The Pennsylvania State University Press, 1973).
54. Sopper, W. E. "Crop Selection and Management Alternatives-Perennials," *Proceedings of a Joint Conference on Recycling Municipal Sludges and Effluents on Land.* Champaign, Ill. July 9-13, 1973, pp. 143-153.
55. Bouwer, H. "Water Quality Aspects of Intermittent Systems Using Secondary Sewage Effluent," presented at the Artificial Groundwater Recharge Conf., Univ. of Reading, England, Sept. 21-24, 1970.
56. McMichael, F. C. and J. E. McKee. "Wastewater Reclamation at Whittier Narrows," Cal. Tech. State Water Quality Bd. Publ. 33, Sacramento (1966
57. Gotaas, H. B., *et al.* "Wastewater Reclamation in Relation to Groundwater Pollution," California State Water Pollution Control Board, Publication No. 6, Sacramento (1953).
58. Baars, J. K. "Experiences in the Netherlands on Contamination of Groundwaters," Proc. 1961 Symp. USHEW, Groundwater Contamination, Robert Taft Eng. Ctr., Cincinnati (1961).
59. Baars, J. K. "Artificial Groundwater Production by Biofiltration in Fine Sandy Soils," *J. Sci. Food Agric.* **8**, 610 (1957).
60. Bendixen, R. W., R. D. Hill, F. T. DuByne and G. G. Robeck. "Cannery Waste Treatment by Spray Irrigation Runoff," *J. Water Poll. Control Fed.* **41**(3), 385 (1969).
61. Law, J. P., Jr. and R. E. Thomas. "Waste Disposal by Overland Flow Spray Irrigation," *Climatology* **22**(2), 73 (1969).
62. Foster, H. B., *et al.* "Nutrient Removal by Effluent Spraying," *JSED-ASCE* **91** (SA6), 1 (1965).
63. Ellis, B. G. and A. E. Erickson. "Movement and Transformation of Various P Compounds in Soil," Soil Sci. Dept., Mich. State Univ. (1969).
64. Ellis, B. G. "The Soil as a Chemical Filter," in *Recycling Treated Municipal Wastewater and Sludge Through Forest and Cropland*, W. E. Sopper and L. T. Kardos, Eds. (University Park, Pa.: The Pennsylvania State University Press, 1973).
65. Bendixen, R. Q., R. D. Hill, W. A. Schwartz and G. G. Robeck. "Ridge and Furrow Liquid Waste Disposal in a Northern Latitude," *JSED-ASCE* **94** (SA1), 147 (1968).
66. Law, J. P., Jr., R. E. Thomas and L. H. Myers. "Cannery Wastewater Treatment by High Rate Spray on Grassland," *J. Water Poll. Control Fed.* **42**(9), 1621 (1970).
67. Lance, J. C. and F. D. Whisler. "Nitrogen Balance in Soil Columns Intermittently Flooded with Secondary Sewage Effluent," *J. Environ. Quality* **1**, 180 (1972).
68. Viets, F. G., Jr., and R. H. Hageman. "Factors Affecting the Accumulation of Nitrate in Soil, Water and Plants," Agriculture Handbook No. 413, (Washington, D.C.: U.S. Department of Agriculture, ARS, 1971).

69. Pennypacker, S. P., W. E. Sopper, and L. T. Kardos. "Renovation of Wastewater Effluent by Irrigation of Forest Land," *J. Water Poll. Control Fed.* **39**(2), 285 (1967).
70. Bouwer, H. "Renovating Secondary Effluent by Groundwater Recharge with Infiltration Basins," in *Recycling Treated Municipal Wastewater and Sludge Through Forest and Cropland*, W. E. Sopper and L. T. Kardos, Eds. (University Park, Pa.: The Pennsylvania State University Press, 1973).
71. Baars, J. K. "Travel of Pollution and Purification en Route in Sandy Soils," *Bull. World Health Org.* **16**, 727 (1957).
72. Biggar, J. W. and R. B. Corey. "Agricultural Drainage and Eutrophication," in *Eutrophication: Causes, Consequences, Correctives.* Proc. of a Symposium, Madison, Wisconsin. (Washington, D.C.: National Academy of Sciences, 1969), pp. 404-445.
73. Driver, C. H., B. G. Hrutfiord, D. E. Spyridakis, E. B. Welch, D. D. Wooldridge, and R. F. Christman. "Assessment of the Effectiveness and Effects of Wastewater Management," U.S. Army Corps of Engineers Wastewater Management Report 72-1 (1972).
74. Blakeslee, P. A. "Monitoring Considerations for Municipal Wastewater Effluent and Sludge Application to the Land," *Proceedings of a Joint Conference on Recycling Municipal Sludges and Effluents on Land.* Champaign, Ill., July 9-13, 1973, pp. 183-198.
75. American Chemical Society. *Cleaning Our Environment: The Chemical Basis for Action.* (Washington, D.C.: American Chemical Society, 1969).
76. Bailey, G. W. and J. L. White. "Soil-Pesticide Relationships," *Agric. Food Chem.* **12**, 324 (1964).
77. Gould, R. F., Ed. *Organic Pesticides in the Environment.* Advances in Chemistry Series No. 60. (Washington, D.C.: American Chemical Society, 1966).
78. Mortvedt, J. J., P. M. Giordano, and W. L. Lindsay, Eds. *Micronutrients in Agriculture.* (Madison, Wisc.: Soil Sci. Soc. Amer., Inc., 1972).
79. Hem, J. D. "Chemistry and Occurrence of Cadmium and Zinc in Surface Water and Groundwater," *Water Resources Res.* **8**(3), 661 (1972).
80. Bowen, H. J. M. *Trace Elements in Biochemistry.* (New York: Academic Press, 1966).
81. Chapman, H. D., Ed. *Diagnostic Criteria for Plants and Soils.* (University of California, Division of Agricultural Sciences, 1966).
82. Toth, S. J. and A. N. Ott. "Characterization of Bottom Sediments: Cation Exchange Capacity and Exchangeable Cation Status," *Environ. Sci. Technol.* **4**(11), 935 (1970).
83. Jenne, E. A. "Controls on Mn, Fe, Co, Ni, Cu, and Zn Concentrations in Soils and Water: The Significant Role of Hydrous Mn and Fe Oxides," in *Trace Inorganics in Water*, R. A. Baker, Ed. Advances in Chemistry Series, 73. (Washington, D.C.: American Chemical Society, 1968), pp. 337-387.
84. Leland, H. V. and N. F. Shimp. "Factors Affecting Distribution of Lead and Other Trace Elements in Sediments of Southern Lake Michigan," in *Trace Metals and Metal-Organic Interactions in Natural Waters*, P. C. Singer, Ed. (Ann Arbor, Mich.: Ann Arbor Science Publishers, Inc., 1973), pp. 89-129.

85. Schnitzer, M. "Reactions Between Organic Matter and Inorganic Soil Constituents," *Trans. Int. Congr. Soil Sci. 9th* **1**, 635 (1968).
86. Wiklander, L. "Cation and Anion Exchange Phenomena," in *Chemistry of the Soil*, E. Bear, Ed. (New York: Reinhold Publishing Corp., 1964).
87. Barrow, N. J. "Comparison of the Adsorption of Molybdate, Sulfate and Phosphate by Soils," *Soil Sci.* **109**, 282 (1970).
88. Sepp, E. "The Use of Sewage for Irrigation–A Literature Review," Bureau of Sanitary Engineering, Calif. State Dept. of Public Health (1971).
89. Thomas, R. E., J. P. Law, Jr., and C. C. Harlin. "Hydrology of Spray Runoff Wastewater Treatment," *J. San. Eng. Div. Amer. Soc. Civil Engr.*, **96**(IR-3, 7538), 289 (1970).
90. Vollenweider, R. A. "Water Management Research, Scientific Fundamentals of the Eutrophication of Lakes and Flowing Waters, with Particular Reference to Nitrogen and Phosphorus as Factors in Eutrophication," Organization for Economic Cooperation and Develop. DAS/CSI/68.27 (1968), 182 pp.
91. Swaine, D. J. "The Trace Element Content of Soils," *Commonwealth Bur. Soil Sci. Tech. Comm.* **48** (1958).
92. Kopp, J. F. and R. C. Kroner. "Trace Metals in Waters of the United States, (Oct. 1, 1962 to Sept. 30, 1967)," (Washington, D.C.: U.S. Dept. of the Interior, FWPCA, 1970), 32 pp.
93. Durum, W. H., J. D. Hem, and S. C. Heidel. "Reconnaissance of Selected Minor Elements in Surface Waters of the United States, October, 1970," *U.S. Geol. Surv. Circ.* **643** (1971), 49 pp.
94. Lazarus, A. L., E. Lorange and J. P. Lodge, Jr. "Lead and Other Metal Ions in United States Precipitation," *Environ. Sci. Technol.* **4**, 55 (1970).
95. Lipman, J. G. and A. B. Conybeare. "Preliminary Note on the Inventory and Balance Sheet of Plant Nutrients in the United States," *New Jersey Agr. Exp. Sta. Bull. 607* (1936), 23 pp.

4

ACCEPTABILITY OF WASTEWATER EFFLUENTS BY SOILS

A. Hayden Ferguson

Professor of Soil Physics
Department of Plant and Soil Science
Montana State University
Bozeman, Montana 59715

INTRODUCTION

For the purposes of this chapter, the title phrase "acceptability of wastewater effluents by soils" will mean that "a soil is acceptable for wastewater if the wastewater can be added to the soil on a semi-continuous basis for long periods of time (many years) to accomplish successfully the removal of all but some minimum acceptable level of undesirable materials from the aquatic stream without itself being destroyed." This definition is intentionally very broad and nonspecific.

The addition of wastewater effluent to soils immediately triggers and/or modifies a whole series of complex chemical, physiochemical, physical, and biological reactions. The consequence of these reactions determines the acceptability of wastewater effluents by soils. In this chapter, we discuss some of the more important reactions and the properties of soils and effluents that control these reactions.

The solid matrix of soils consists of sand, silt, clay, and organic matter. Because of their small relative surface area, the sand and silt fractions are essentially nonreactive. They provide a relatively rigid framework containing the clay and organic matter but by themselves function largely only as a physical filter. Conversely, the clays and organic materials are extremely reactive, thus determining to a major degree the acceptability of effluents by soils.

ORGANIC MATTER

Organic matter occurs in a spectrum of forms from raw sticks to extremely complex organic compounds that are generally lumped under the name humus. Humus is extremely active material. It has a tremendous capacity to absorb water, it has many reactive sites that have the ability to fix in exchangeable form both anions and cations, and it forms many relatively stable complexes with many cations and anions. Moreover, humus markedly modifies the reaction of clays. The quantity of organic matter in soils is usually small, averaging from about 2 to 10%. Organic matter has typical cation exchange capacity of about 200 me/100 g.

CLAYS

In most soils, clay is the most active component. Clays are generally defined as those mineral particles that have an effective diameter of less than 2 microns. Particles of this size have tremendous surface areas relative to their masses; even in sandy soils over 95% of the total surface area is associated with the clay.

Clays are generally crystalline, made up of the anions oxygen and hydroxyl tied together with the cations aluminum and silicon. Thus, at the edges, broken bonds associated with both the anions and the cations exist. Moreover, many of the silicate clays have undergone isomorphic substitution, a process whereby a cation of lower valence has been substituted for aluminum or silicon. This results in the clay particles having an internal negative charge, which is usually described in terms of cation exchange capacity.

There are many types of silicate clays but only two will be discussed here: kaolinite, which tends to be dominant in humid areas, is relatively large-sized and has a low cation exchange capacity of 5-15 me/100 g, and montmorillonite, which tends to be dominant in more arid areas, is small in size and has a large cation exchange capacity of about 100 me/100 g. The charge associated with the broken bonds and isomorphic substitution together with the large surface area make clays the very active substances they are.

Reaction of Clays

Figure 4.1 is a schematic of a clay particle, showing the layers previously described. The horizontal dimensions of this clay particle are vastly greater than the vertical. The dashes represent sites of negative charges which must be neutralized by cations, and the small pluses represent cations. In kaolinite the regions denoted by the 1 are essentially absent since the

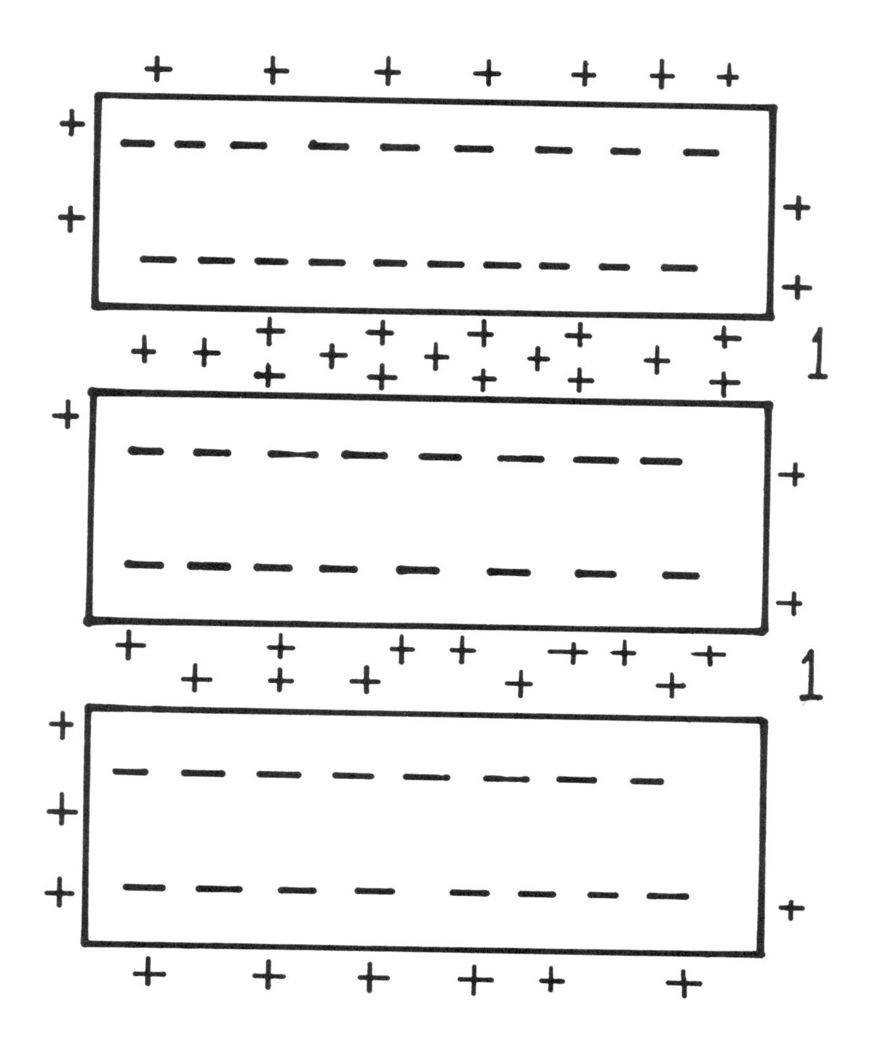

Figure 4.1. A clay particle consisting of three unit layers. The dashes are negative charges in the crystalline system. The pluses are positive counter ions. In kaolinite the regions denoted by 1 are fixed and contain no ions or water; in montmorillonite these regions contain exchangeable ions and water can easily move in and out.

individual units are held firmly together. In montmorillonite, the regions denoted by the 1 between the individual units contain water and cations. The unit can expand or contract, keeping its relative position, as water moves in or out of region 1.

The strength of attraction of cations to the clay particle is largely dependent upon two factors: valence and size (which includes the water hull). High valence cations are attracted with greater force than low valence cations, and small cations are attracted more strongly than large, assuming identical valences. The importance of strength of attraction is discussed later.

The tendency of a mass of particles to swell and shrink is highly dependent upon particle size: small particles tend to swell more. All clays tend to swell and shrink but montmorillonite clays tend to swell and shrink more than kaolinite clays. However, the magnitude of shrink-swell reactions of clays is less related to size than to the clay-cation interactions.

Figure 4.2 shows the dimensional distribution of cations away from one side of a clay particle in a wet system. The concentration of cations is greatest near the clay and decreases with distance. This distribution is established by a combination of coulombic attractive forces and concentration-dependent repulsive forces, and it is commonly denoted as the diffuse double

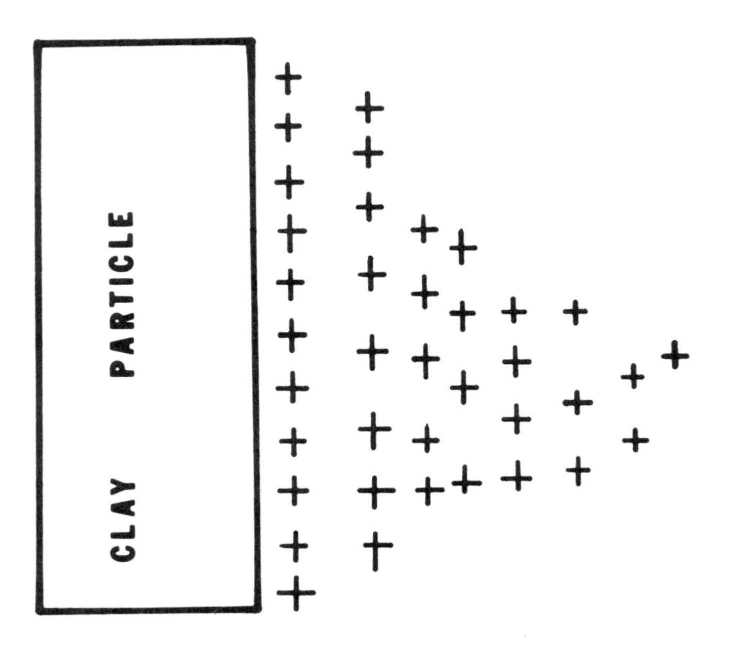

Figure 4.2. One-dimensional distribution of exchangeable cations away from one side of a clay particle.

layer. The thickness of the diffuse layer in soils is of major importance in controlling the tendency of a soil to swell, to conduct water, to support foundations and many other important phenomena. In general, the thicker the double layer the greater the problems in dealing with and using the soil. This phenomenon can be graphed as shown in Figure 4.3. The lower part of

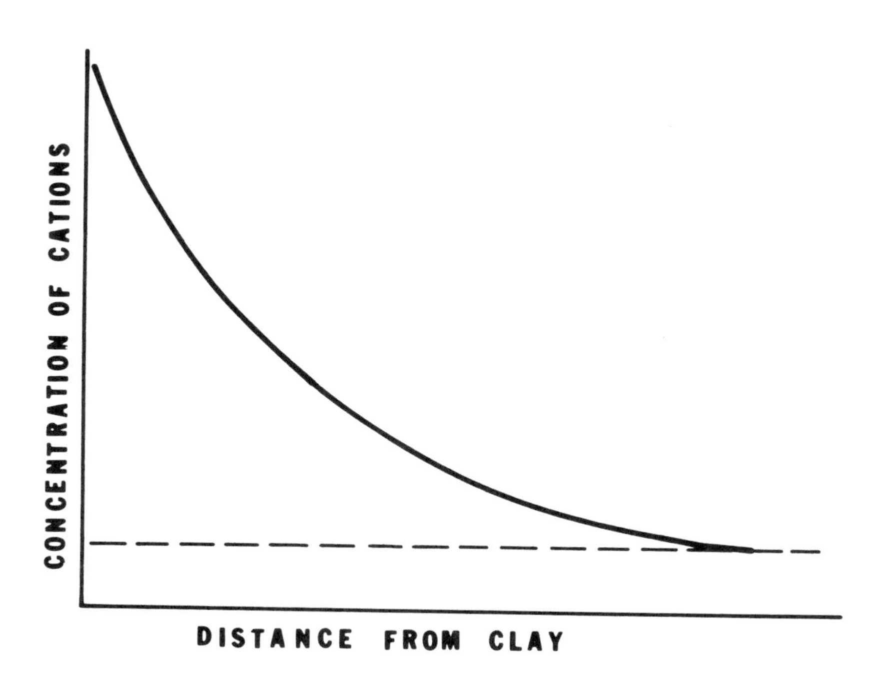

Figure 4.3. Cation concentration as a function of distance away from a clay particle. The dashed line represents the concentration of cations in the soil solution beyond the influence of the clay.

the curve, designated by the dashed line, is the original concentration of cations in the soil solution some distance from a soil particle. For the cations calcium and sodium this graph would be as shown in Figure 4.4. The calcium is small and has a valence of 2, the sodium is large and has a valence of 1. Note that the double layer is much thicker with the sodium ion than with the calcium ion.

Another factor that influences the distribution of cations away from a clay particle is the concentration of cations in the external soil solution (*external* refers to the concentration of cations in the middle of big pores

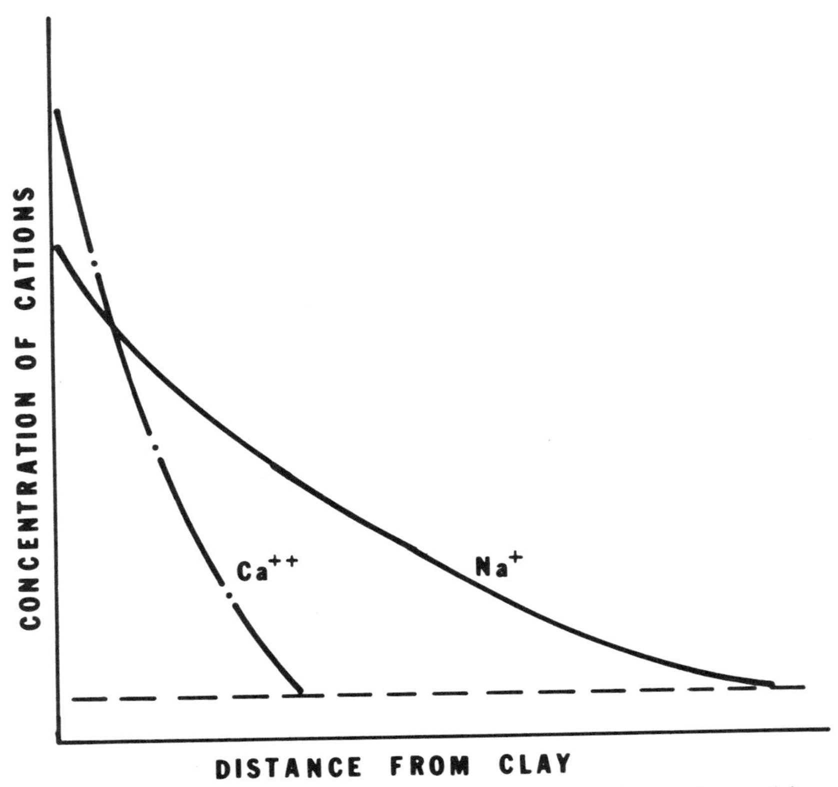

Figure 4.4. Distribution of calcium and sodium ions away from a clay particle.

away from the influence of clay particles). This influence is shown in Figure 4.5. Note that with the low concentration the clay particle influences the cations for a much greater distance.

Consider, now, how these factors influence the swelling of soils. The solid lines in Figure 4.6 show two adjacent clay particles in a stable position. In an intermediate salt concentration, 2 mmhos/cm, with calcium as the dominant cation, the swarm of cations associated with each particle is at a concentration equal to the concentration of the external solution midway between the two clay particles. If this system were suddenly changed so that sodium became the dominant cation, the diagram would immediately change to that shown by the dashed lines. The concentration of cations at the midpoint between the clay particles would suddenly become higher than the concentration of cations in the external solution. This would cause a difference in the osmotic pressure between the clay

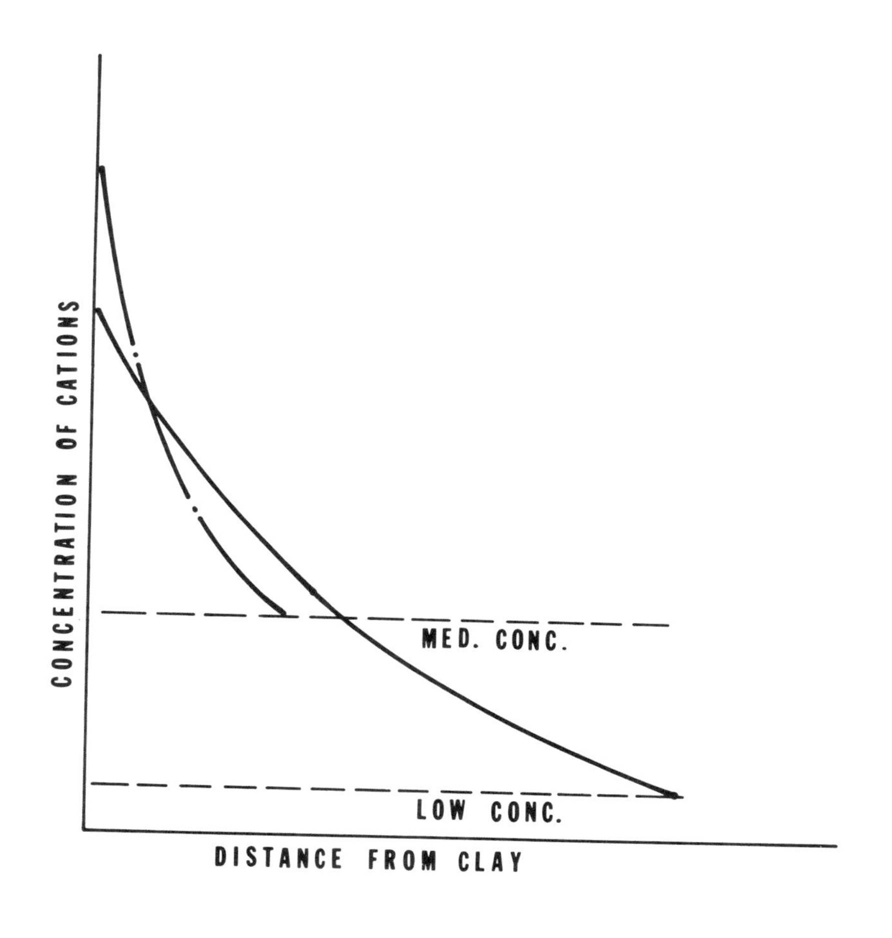

Figure 4.5. The influence of salt concentration in the soil solution on the concentration-distance curve.

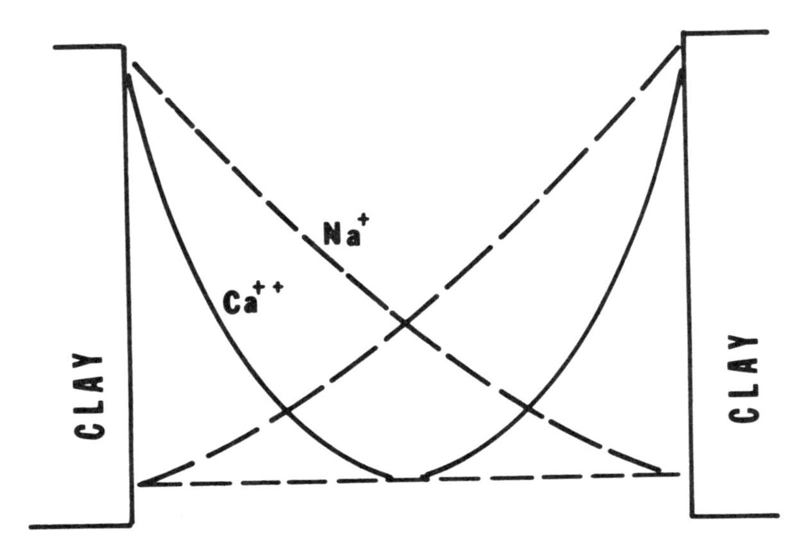

Figure 4.6. The solid line represents a stable configuration with calcium as the exchangeable ion. If the calcium were suddenly replaced with sodium, the dashed line represents the instantaneous double layer. In this case the osmotic potential difference will cause water movement into the area between the clay particles, causing swelling.

particles and the external solution, which would pull water from large pores into the zone between the particles. Stability would be obtained in the case of the sodium only when this water movement had forced the particles apart to the point that the midpoint double layer concentration equaled that of the external solution.

Swelling, then, is promoted by a dominance of sodium on the exchange complex, especially under conditions of low salt concentrations in the soil solution, and it is retarded by a dominance of calcium and/or high soluble salts. A convenient method of describing the influence of the exchangeable ions in the soil system or in water added to the soil is to consider the ratio of sodium to divalent ions. This is usually described in terms of the sodium adsorption ratio (SAR), which is defined as

$$\mathrm{SAR} = \frac{[\mathrm{Na+}]}{\sqrt{\frac{[\mathrm{Ca++}]\ [\mathrm{Mg++}]}{2}}}$$

As a general rule, an SAR of 15 or greater is considered unacceptable. However, there is a difference depending on the dominant clay type; with kaolinite, the SAR can be as high as 20 before serious swelling problems occur, but SAR's of 8-10 will cause serious swelling problems with soils in which montmorillonite is dominant.

SOLUBILITY REACTIONS

A major factor determining the fate of materials that may be added to soils in sewage effluents is the solubility of the reaction products formed. In this section, we will discuss the solubility-insolubility of some materials commonly found in wastewater effluents when they are added to soils.

Nitrogen

Most of the nitrogen in wastewater exists as NO_3^- or NH_4^+, with the exception being a small amount of nitrogen in soluble organic complexes. As such, the nitrogen remains soluble when added to the soil. Almost the only activity that could result in insoluble forms of nitrogen in the soil would be the uptake of the nitrogen by a biological organism. Both NO_3^- and NH_4^+ are taken up readily by most organisms. In soluble form, NO_3^- and NH_4^+ act very differently in the soil-water system, but these differences will be discussed later.

Phosphorus

Phosphorus forms highly insoluble compounds with iron, aluminum, and calcium. These are common elements of soils and their activity in any soil is to a large degree a function of pH. At a pH below 6, iron and aluminum become active, and insoluble compounds such as strengite, $Fe(H_2PO_4)(OH)_2$, and variscite, $Al(H_2PO_4)(OH)_2$, with solubility products of about 10^{-35} and 10^{-31} are formed. At a pH above 6.5 to 7, phosphorus reacts with calcium to form, finally, octocalcium phosphate, $Ca_4H(PO_4)_3 \cdot 3H_2O$, which has a solubility product of about 10^{-47}. Thus, at pH's below 6 and above 7, phosphorus from wastewater is removed almost quantitatively from solution by most soils.

Besides these precipitation reactions, other reactions, generally termed fixation reactions, occur that remove phosphorus from solution. Perhaps the most important of these is the fixation of phosphorus on the silicate clays. The aluminum hydroxide at the broken edges of these clays has a great ability to fix phosphorus, and phosphorus is also fixed on the silicate clays through some sort of a clay-Ca-P linkage. Another fixation reaction occurs in soils containing particles of calcium carbonate (soils with pH values above 8 usually contain calcium carbonate) in that phosphorus is adsorbed on the calcium carbonate particles.

At pH values of less than 6 and greater than 7, almost complete retention of phosphorus can be expected when sewage effluent is added as irrigation water to most soils. In soils with a pH between 6 and 7, phosphorus tends to remain more soluble, but in silicate clay soils retention should be nearly complete under most systems of addition.

Heavy Metals

Although data are scarce, it appears that in most soils, heavy metals are retained in the soil by adsorption on clays and organic matter. This would be especially true in soils with a pH of 7 or above. Most heavy metals are toxic in high concentrations, but it does not appear that the heavy metal concentration found in most domestic sewage effluents would create toxicity problems for plants.

EXCHANGE REACTIONS

Exchange reactions are of real importance in determining the fate of ions added to soils in sewage effluents. In most soils, the dominant charge of the soil complex is negative. Thus cations in the effluent tend to be fixed in exchangeable form on the clays, and organic matter and anions tend to be repelled from these surfaces and forced out into the "large" channels. This results in very significant differences in the rate at which cations and anions move through the soil systems.

Anions tend to move with the water; in fact, they generally accumulate near the head of any wetting front of water moving through the soil. On the other hand, cations that are fixed in exchangeable form tend to remain in place until replaced by another cation. The tendency of a soil to retain various cations in exchangeable form depends on several factors, with valence and degree of hydration being among the more important. This tendency usually follows the general order $Li^+ < Na^+ < NH_4^+ < K^+ < Rb^+ < Cs^+ < Mg^{++} < Ca^{++} < Sr^{++} < Ba^{++}$.

Reactions somewhat different from those described above may occur in soils of humid regions such as southeastern United States or Hawaii. Soils of these areas tend to have higher anion exchange capacities and their exchange reactions are both pH and concentration dependent.[1] Thus, care must be taken in extending these generalizations about temperate region soils to humid region soils.

The soil solution phase tends to maintain some equilibrium dependent upon both the effluent and the fixed soil phase. Thus, the amount and composition of soluble salts in the soil solution will be proportional to those of the effluent. After continued application of effluent, the soil solution should approach that of the effluent. An exception to this would be the occurrence of evapotranspiration without adequate leaching. Under this condition, the soil solution would become highly concentrated and eventually cause the death of all plants. Leaching requirements to prevent harmful salt buildup are described in U.S. Department of Agriculture Handbook No. 60, available from the Superintendent of Documents, Washington, D.C.

Breakthrough curves for the addition of magnesium and chloride to a sand soil are shown in Figure 4.7.[2] These data show the rapid movement of the anion relative to the cation. They also show that eventually the soil solution takes on the characteristics of the effluent.

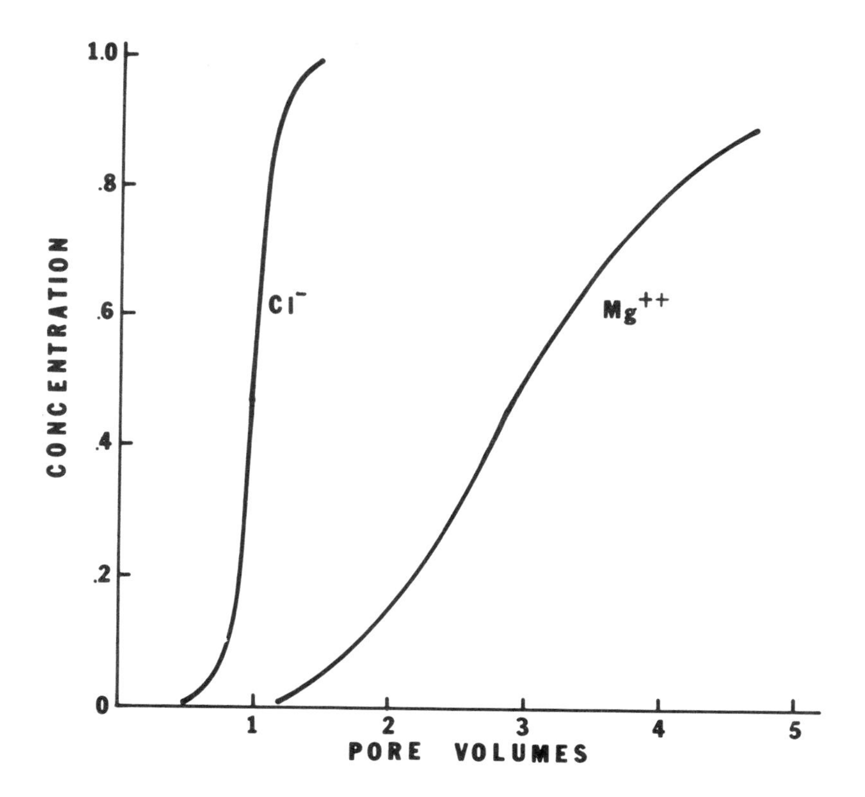

Figure 4.7. Breakthrough curves showing the relative concentration of Cl^- and Mg^{++} in the percolate when water containing only these ions is added to a sandy soil. The anion moves readily through the soil, the cation is retarded by exchange reactions. Eventually the Mg will reach 1.

The magnitude of these exchange reactions depends upon the cation exchange capacity (CEC) of the soil, in turn a function of the type and quantity of clay and organic matter. Typical soils may have the following CEC values: sandy soils, 2-4 me/100 g; sandy loams, 4-10 me/100 g; loams, 10-15 me/100 g; silt loams, 10-25 me/100 g; clay and clay loams, 20-60 me/100 g. These kinds of values can be significant; for example, a soil with a CEC of 25 me/100 g containing Ca^{++} as the dominant exchangeable ion might contain about 4 tons of exchangeable Ca^{++} in a 6-inch layer one acre in area.[3]

OXIDATION-REDUCTION

Many components of the system may undergo oxidation-reduction reactions when wastewater is added to soils. These reactions are especially important in the case of nitrogen because it is potentially a serious pollutant in wastewater and its fate in the system is highly dependent on its state of oxidation. In aerated systems the nitrogen tends to form nitrate and therefore tends to move with the water. In nonaerated systems, the nitrogen tends toward reduced status—NH_4^+, NH_3 and N_2. As NH_4^+ it acts as any cation; as NH_3 or N_2 little downward movement with the water occurs and the nitrogen tends to be lost from the system as gas.

SOIL WATER MOVEMENT

According to Bouwer[4] land disposal of wastewater can occur as (1) low-rate systems for irrigation and water renovation with applications of about 1 to 4 inches per week or (2) high-rate systems with effluent renovation as the main objective and application rates as high as several feet of water per day [sic]. The type of water movement is very different in these two systems but soil water movement is of major importance in both.

Soil water movement, both infiltration and percolation, is highly dependent upon pore size distribution (not necessarily porosity) of the soil. Water moves readily through large pores and as the pore size decreases, the resistance to water flow increases. Pore size within soils is related to the texture and the structure. Texture by itself is a fair but not foolproof indicator of general pore size. Sandy soils have relatively large pores, thus transmitting water rapidly, and clay soils have small pores with relatively slow water transmission properties. Superimposed upon texture is structure, or the tendency of the clays to be flocculated into relatively large secondary aggregates or to be dispersed so that no stable secondary aggregates exist. This phenomenon is, to a major degree, controlled by the factors discussed earlier under reactions of clays. Systems that result in a high degree of swelling (dispersion), which have high SAR and relatively low electrolyte concentration, will result in small pores, even in sandy soils. Conversely, high permeability can be maintained in soils high in clay provided the clay remains flocculated. Even under conditions conducive to flocculation, the physical action of sprinkler irrigation can cause dispersion of surface soil aggregates and a reduction in the infiltration rate. Other factors that may decrease pore size in soils are clogging by suspended solids, microbial by-products and/or chemical precipitates such as ferrous sulfide. The latter two are most apt to occur under anaerobic conditions, which are often produced by high-rate systems.

The high-rate system, in which large depths of water are ponded in basins, is mainly a saturated flow system. As such, water flow can be described by the Darcy equation:

$$\frac{Q}{TA} = K \frac{\Delta h}{\Delta l}$$

where Q, T, A, h and l are volume of flow, time, area, head, and depth of saturated soil, respectively. K is the hydraulic conductivity and describes the ease of water flow. Soils in which K cannot be sustained at about 1 in./hr or higher are questionable for high-rate systems. Such rates can, in general, only be maintained on soils of sandy or coarse texture in which any clay is flocculated and where the management system prevents clogging. Soils with very high K values, perhaps greater than 4 in./hr, would not result in adequate effluent renovation. Even with ideal K values, *e.g.*, 2 in./hr, the soil depth must be about 10 feet to obtain satisfactory renovation of effluent.

It should be noted that K tends to decrease with time in almost all soils because the reactions that reduce pore size are time-dependent. Also, K is very much a function of the water applied. Short term experiments (2-3 days) with water of a different soluble salt content or SAR value than the sewage effluent are very apt to furnish completely misleading data.

In low-rate irrigation systems water movement is as unsaturated flow. The only important exception, providing the soil does not swell and become impermeable, is at the surface when the application rate exceeds the infiltration rate. Here water movement occurs, not only because of the gravitational potential gradient, but also because of other potential gradients in the soil. The movement is very much a function of the water content of the soil: infiltration rate decreases as the soil water content increases but, for a particular potential gradient, the rate of water movement increases with increasing water content.

Water application rates should not exceed the soil infiltration rate except perhaps on level areas where runoff will not occur. Otherwise, serious pollution problems might arise. Moreover, many crops are sensitive to the poor aeration that accompanies high application rates. Alfalfa is an example, and since the crop is an important part of the renovation process it should not be damaged. A minimum soil infiltration capacity is difficult to define since it would vary with the land area available and the quantity of effluent to be disposed of. However, soils with a saturated hydraulic conductivity of less than about 0.04 in./hr are questionable. As stated before, these data should be based on a run of several days duration using water of an SAR and salt content similar to the sewage effluent.

The total periodic (weekly) water application can be tied to estimates of evapotranspiration and the soil water-holding capacity. By adding this amount, plus an additional amount of roughly 1 in./week, the crop water needs can be supplied and the excess renovated water that moves beyond the root zone supplied to the water table. With the exception of nitrate and perhaps a few other substances such as boron, this system should result in more complete renovation than the high-rate system. Soils with much higher surface areas can be utilized. The flow rates are slower and a greater proportion of the total flow occurs in smaller pores. Therefore, adsorption of components such as heavy metals and phosphorus can be much more complete.

Nitrogen removal can be very complete in this system; the aerobic system results in the transformation of organic nitrogen to NH_4^+ and NO_3^-, and these forms, especially nitrate, are readily taken up by plants and microorganisms. However, the fact that nitrogen tends to be in nitrate form in this system can lead to serious pollution problems. Rapid water movement to beyond the rooting depth, which occurs with excess water application to permeable soils (even clay soils can fit this description), will lead to nitrate enrichment of groundwater. Once the nitrate has moved beyond the root zone in this system, it is potentially a serious pollutant. Moreover, since the nitrate tends to be accumulated near the wetting front, the rapid movement of only a small amount of water below the depth can be detrimental.

SITE SELECTION

The site requirements for high-rate systems, specifically the necessity of very highly sustained permeability, may seriously limit the location of these systems. The site requirements for irrigation systems, however, are much less restrictive. Thus, these systems have more general application, although minimum site requirements must be met for irrigation systems. Each site should be thoroughly investigated to establish suitability as a disposal area. Among the more important parameters to be investigated are:

(1) Soil depth: should be not less than 4-5 feet over gravel and/or the water table. If an impermeable zone underlies the area, the depth should be much greater than this, probably not less than 15-20 ft.
(2) Water table fluctuations: water tables that rise to less than 4 feet from the surface may present serious hazards.
(3) Soil texture: should be indicated by depths, and should include reliable estimates of clay type.
(4) Hydraulic conductivity: should be determined on all distinctly different horizons. Marked changes in hydraulic conductivity will result in marked changes in water flow properties.
(5) Chemical properties: these include pH, SAR, and soluble salts as a minimum and should be made on each soil horizon.
(6) Water quality: effluent constituents should be determined.

The integrated information from these measured parameters should allow for the accurate prediction of the suitability of a site.

ADDITIONAL CONSIDERATIONS

Many local considerations must be investigated. This is especially important in terms of effluent constituents. For example, in areas where water hardness causes extensive use of water softeners one might expect "detrimental" effluent SAR. Also, heavy metal concentration may depend significantly on the geology and the economic activity of an area. In areas where heavy metal concentrations are suspected, thorough research is suggested before investing in an irrigation system. Some effects are rather subtle. For example, in areas of rather high but nontoxic levels of boron, the morning flush of boron from shaving cream added to the water might harm crops.

REFERENCES

1. Van Raij, B. and M. Peech. "Electrochemical Properties of Some Oxisols and Alfisols of the Tropics," *Soil Sci. Soc. Amer. Proc.* **36**, 587 (1972).
2. Bigger, J. W. and D. R. Nielsen. "Miscible Displacement and Leaching Phenomenon," *Irrigation of Agricultural Lands*, Agronomy No. 11 (Madison, Wisc.: American Society of Agronomy, 1967), p. 254.
3. Murrmann, R. P. and F. R. Koutz. "Role of Soil Chemical Processes in Reclamation of Waste Water Applied to Land," *Waste Water Management by Disposal on the Land*, Special Report 171 (Hanover, N.H.: Cold Regions Research and Engineering Lab., 1972), p. 48.
4. Bouwer, H. "Returning Wastes to the Land, A New Role for Agriculture," *J. Soil Water Cons.* 23, 164 (1968).

ADDITIONAL REFERENCES

Methods of Soil Analysis: Physical, Agronomy 9, Part 1. (Madison, Wisc.: American Society of Agronomy).

Methods of Soil Analysis: Chemical, Agronomy 9, Part 2. (Madison, Wisc.: American Society of Agronomy).

Irrigation of Agricultural Lands, Agronomy 11 (Madison, Wisc.: American Society of Agronomy).

5

PUBLIC HEALTH IMPLICATIONS OF THE APPLICATION OF WASTEWATERS TO LAND

A. W. Hoadley and S. M. Goyal

Associate Professor and Postdoctoral Fellow
School of Civil Engineering
Georgia Institute of Technology
Atlanta, Georgia 30332

INTRODUCTION

Questions of the public health as it may be related to the application of wastewaters and sludges to the land have been considered by a number of authors. Shuval[1] has reviewed the broad health factors that should be considered in planning for the utilization of wastewaters. More recently Benarde,[2] in a thorough appraisal of infectious disease related to the application of sewage effluents to the land, concluded that as a matter of good public health practice, application to the land would properly include secondary treatment followed by chlorination. However, Bernarde concluded also that in the United States, the potential hazard associated with the application of wastewaters to the land is low, and in fact, is less than that associated with their discharge to surface waters. Indeed, many drinking water supplies in the United States draw their water from polluted sources. However, there do exist certain potential risks to the public health associated with the use of wastewaters for irrigation or the application of sludges to the land.

It is not the intent in the present discussion to reiterate what has been said by the above authors, and the reader is referred to their excellent papers for their views. Neither is it our intent, in presenting a discussion of potential health considerations, to discourage the application of wastewaters and sludges to the land. Rather, it is our wish to elaborate on potentially important pathogens and hazards associated with their dissemination, to identify and put into perspective potential routes of infection that have not been explored extensively in the above reviews, and to call attention to needed surveillance and controls.

SOME EPIDEMIOLOGIC CONSIDERATIONS

A point that should be borne in mind throughout the succeeding discussion relates to the low fecal carriage rates of agents of infectious disease and the low background of enteric disease in the United States today. Because the level of carriers in the population is low, it has been suggested that the potential for the spread of the infectious agents as a result of the application of fecal wastes to the land is limited.[2] But levels of enteric disease in the United States today are low primarily because we have controlled the spread of enteric pathogens by means of sanitary engineering works and the practice of good sanitation and personal hygiene. In recent years the carrier rate has been limited through immunization and antibiotic therapy, but the major progress in the control of these diseases had been accomplished through sanitation of the environment prior to the introduction of measures for the reduction of susceptibility to disease and prior to control of clinical disease. Thus, as exposure of the population to even small numbers of pathogens increases as a consequence of application of wastewaters and sludges to the land, can the levels of fecal carriage and of disease in the population be expected to do likewise?

That populations in the United States are highly susceptible to enteric disease where contact with reservoirs exists is demonstrated amply by (1) the association of human salmonellosis with contaminated food, especially poultry[3] processed at plants handling from 75,000 to 85,000 birds per day, providing ample opportunity for cross-contamination among birds and flocks,[4] (2) the association of an estimated 280,000 cases of salmonellosis per year among children with pet turtles raised in contaminated ponds,[5-8] and (3) shigellosis and hepatitis A associated with daycare centers.[9,10] Weismann *et al.*[9] have suggested that among those at risk from outbreaks of shigellosis at daycare centers is the community as a whole, and that the increasing incidence of *Shigella sonnei* infections evident in several large cities may reflect the impact of daycare-associated shigellosis on urban communities.

The application of wastes to the land may complete cycles of infection not presently existing, or functioning at a low level, that could lead to the establishment of new reservoirs or increases in the levels of contamination in those already existing. The inherent dangers of the system are evident from epidemics occurring from time to time at integrated poultry operations where offal is reused in feed, which in turn is utilized at the same operation. (for example, see Ref. 11). The rapid spread of *Salmonella agona* in the human population and its emergence as a predominant serotype, not only in the United States, but in Israel, the Netherlands, and the United Kingdom as well,[12] provides an instructive model describing the enormous potential that exists for the dissemination of enteric pathogens in the human community. Since it is generally accepted that numbers of salmonellae and their serotypes present in sewage **reflect** levels and agents of disease in the community, respectively (for example, see Ref. 13), increases in the rates of infection may be expected to increase the contamination of the environment and the potential hazard.

Often the lack of epidemiologic evidence of disease outbreaks attributable to wastes applied to the land has been cited as further evidence that the threat to human health is minimal (for example, see Ref. 2). In 1956 and 1957, reclaimed wastewater was used for public water supply in Chanute, Kansas, as an emergency measure for a limited time with no apparent adverse health effects.[14] Reclaimed effluents have been used also in Lyndon, Kansas, and Ottumwa, Iowa.[15] Renovated wastewater has been used successfully as a supplement to potable supplies at Windhoek, South Africa. There also appears to be no record either in the United Kingdom or in the United States of an epidemic definitely attributable to the handling of sewage sludge or to the consumption of crops grown on land fertilized with sludge. Nelson *et al.*[16] stated that no outbreak of disease among animals had been traced to sewage on a sewage farm during its 100 years in existence. Merrel and Ward,[17] during investigations of the Santee Project in California, demonstrated that viruses could be detected in almost all samples of raw and primary sewage and from activated sludge effluent. They also could be isolated from 9% of samples of chlorinated pond effluent, but never from effluent after percolation through 200 feet (61 meters) of soil. Indeed, California authorities state that such water may be used not only for irrigation but for recreational purposes as well. Furthermore, a Colorado golf course is spray-irrigated with wastewater originating from a tuberculosis hospital without apparent hazard to health.[18] On the other hand, a recent outbreak of shigellosis among persons swimming in the Mississippi River near Dubuque, Iowa,[19] and a number of known outbreaks among animals exposed to sewage (see discussion below) indicate that the lack of evidence commonly cited may be more apparent than real.

It should be recognized that the reporting of cases of enteric disease, including food and waterborne disease, is notoriously poor,[3,20] and such disease related to the land application of wastes (or indeed disposal to surface waters) might not be recognized without intensive surveillance.

Furthermore, there exist few major land application projects in the United States or other industrialized countries where rates of enteric disease are low and levels of sanitation are high. Where projects do exist, effluent quality is probably well controlled. So there has been little opportunity for observation of outbreaks or increased rates of disease.

The major potential hazard exists where there occurs a lapse in the control of the process, as in the Keene, N.H.,[21] and Zermatt, Switzerland[22] typhoid fever outbreaks. This is particularly true if epidemic strains of the more serious enteric pathogens such as the El Tor cholera vibrio, *Salmonella typhi*, or *Shigella dysenteriae* are imported into the country from abroad. Barker[23] has stressed the effectiveness of sanitation in the United States, which has led to the failure of such epidemic strains to become established. That epidemic cases might arise under relaxed vigilance is suggested by the outbreak of cholera El Tor in Jerusalem in 1970, which was attributed to the use of sewage for irrigation of vegetables sold in the city.[24] Thus, surveillance of disease and carrier rates in human and animal populations at risk, as well as surveillance of the environment, should be undertaken as part of any land application system.

PATHOGENS IN SEWAGE AND SLUDGE

That a wide variety of parasites, pathogenic bacteria, and viruses occur in sewage needs little emphasis here. However, the sanitary character of sewage varies from community to community depending upon urbanization, population density, and sanitary habits. Bacterial pathogens that may be present in sewage include *Salmonella typhi,*[25-28] paratyphoid and other *Salmonella* serotypes,[13,27,29-34] *Pseudomonas aeruginosa,*[35] mycobacteria,[36-42] and *Clostridium perfringens.*[43-45] The only viral disease definitely known to be waterborne is hepatitis A (infectious hepatitis). Both water- and foodborne outbreaks of hepatitis A are well documented, and sewage-contaminated water often has been incriminated as the source of the disease. On the other hand, numerous viruses are excreted by man[46,47] including poliovirus (3 types), coxsackievirus (25 types), echovirus (25 types), reovirus (3 types), and adenovirus (33 types). Various parasites reported from wastewater include *Entamoeba histolytica,*[14] *Entamoeba coli,*[48-50] schistosomes,[51] *Naegleria,*[52] various nematodes,[53-57] and fungi.[58]

While investigations of shigellae and *Vibrio cholerae* in sewage are not reported often in the literature, they are most certainly present where carriers exist in the population served. The incidence of shigellosis in the United States reported to the Center for Disease Control in 1958[59] was 4.6 cases per 100,000 population and is increasing steadily. The actual incidence is certainly very much higher than the reported incidence, perhaps by a factor of 100. Reported rates of infection are higher among high-risk groups, including Indians on reservations (212.5 per 100,000) and patients in mental institutions (1,350 per 100,000). Furthermore, the number of subclinical cases exceeds the number of clinical cases. While short term excretion of shigellae lasting for periods up to several weeks is usual, long term excretion of 10^4 to 10^8 shigellae per gm of feces lasting for nearly 18 months has been reported.[60] Outbreaks of waterborne shigellosis attributable to contamination of water supplies with sewage add support to the assumption that shigellae are present in sewage.[61]

Excretors of the cholera vibrio, of course, would be expected only rarely in the United States. Cholera infections might be anticipated only among recent world travellers, in whom the carrier state probably would never exceed 30 days.[62] Carrier rates reported in India vary from about 1.5% to 7% in the general population and from 4.5% to 33.8% among close contacts of cases.[62,63]

In addition to pathogens in wastes of human and animal origin, concern has been expressed in recent years over antibiotic resistant *Enterobacteriaceae*, which have become widespread as a result of the extensive clinical use of antibiotics and their incorporation into animal feeds. In large measure, resistance is carried by extrachromosomal elements and is transferable between cells, not only of the same species or genus but among genera as well. The true public health significance of bacteria carrying transfer factors in the environment is not at all clear. There have been few incidents in which danger to human health has been proven, although the resistance of the epidemic strain of *S. typhi* in Mexico to chloramphenicol mediated by a resistance transfer factor (and the resistance of some strains to ampicillin as well) illustrates the potential problem that transferable resistance may create. Because there exists little evidence of danger, Jukes[64,65] considers the risks of the widespread use of antibiotics in feeds minor in comparison to the benefits.

On the other hand, other authors, including Anderson,[66] express more concern. Walton[67] has called for continued surveillance so that potential hazards may be dealt with rationally, and Grabow *et al.*[68] have recommended a reexamination of bacterial water quality standards since water may play an important role in spreading transferable resistance. That resistance transfer factors are carried by 1-3% of isolates of *Enterobacteriaceae* from raw and treated sewage has been demonstrated.[69,70] More recently Grabow and Prozesky[71] demonstrated transferable resistance in 26% of coliforms from hospital wastes, and Fontaine and Hoadley (unpublished) demonstrated transfer by 56.2% of fecal coliforms from hospital wastes and up to 74.4% of fecal coliforms isolated from combined hospital and community wastes.

Infective doses of most enteric bacterial pathogens are relatively high. For instance, approximately 10^8 enteropathogenic *Escherichia coli* or *V. cholerae* cells must be consumed by healthy male volunteers to produce disease in a "significant" proportion of subjects. Approximately 10^5 *Salmonella* cells (including *S. typhi*) are required to cause disease, but only 10 to 100 *Shigella* cells are necessary to cause dysentery.[72] Children, sick people, and old people are more susceptible, however, and fewer cells are needed to produce disease. Thus, only 15,000 *Salmonella cubana* cells caused disease in hospitalized patients.[73] Furthermore, prior treatment with 1 gm of oral streptomycin was found by Hornick *et al.*[74] to reduce the infective dose of *S. typhi* from 100,000 to 1,000 cells. But it must be recognized that cross-contamination by smaller numbers of cells entering the home on food, followed by growth in an appropriate medium such as gravy, may lead to the production of large numbers of cells that can and frequently do cause disease.

Counts of microbial pathogens in sewage vary considerably, depending upon rates of disease in the contributing population. Thus, many parasites and the cholera vibrio might be present in sewage in the United States only temporarily after a rare traveller carrying the organisms has joined the community. Other organisms, such as *Entamoeba histolytica*, might be expected only in low numbers, if at all.[48] Some bacterial pathogens, including *P. aeruginosa*, are consistently present in sewage. Fecal carriage of this species occurs in approximately 11% of healthy adults, and populations in excess of 10^5 bacteria per 100 ml are common in raw sewage. Populations in excess of 10^6 bacteria per 100 ml have been reported in sewage in Germany and in hospital wastes in South Africa and the United States.[75] Populations of other bacterial pathogens, such as salmonellae, fluctuate with rates of disease in the community.[13] Kehr and Butterfield[76] estimated that about six typhoid organisms were present in sewage for every million coliforms. But numbers probably are highly variable. Cheng *et al.*[77] demonstrated that populations of nontyphoid salmonellae in raw sewage varied from about 110 to 11,000 per 100 ml. High populations occurred most consistently during August and September when the reported isolations from human sources normally are greatest.[78] Since the number of isolations of shigellae also varies seasonally, the numbers of shigellae in sewage undoubtedly fluctuate in a similar manner.

Enteric viruses have been demonstrated in raw sewage in concentrations of 1 to 10 virus particles per ml in various countries.[79] The concentrations of virus particles discharged in domestic wastes, as those of most enteric bacterial pathogens, are determined by the number of recognized cases of viral disease and the number of unrecognized carriers in the community.[80] Kollins[81] estimated that 500 virus particles were present in 100 ml of sewage during the summer. Kelly and Sanderson[82] reported 4000 virus particles per liter during the warmer months of the year and 200 particles per liter during the colder months. More recently, Wolf[83] demonstrated 5 virus particles per 100 ml of sewage in cold weather months. Shuval[84] made quantitative virus assays of raw wastewater in five cities in Israel and demonstrated an average of 1050 virus particles per liter, although individual determinations varied from 5 to 11,000 virus units per liter. Shuval was able to demonstrate 1 virus particle per 10^6 to 10^7 coliform organisms. In contrast, Clarke and Kabler[85] demonstrated 1 virus particle per 65,000 coliforms, and a ratio of 1 virus particle per 50,000 to 100,000 coliforms may be usual.[86]

Because sewage contains pathogenic microorganisms and viruses, it seems reasonable that settled sludge should contain large numbers. Primary settled sludges contain large populations of microbial pathogens[87,88] and may be highly contaminated with viruses as well.[89] Sludge produced by flocculation of sewage solids[90] and activated sludge also may contain large numbers of virus particles. Heukelekian and Albanese[91] demonstrated that concentrations of tubercle bacilli in settled sludge were 13 times greater than in the supernatant liquor. That salmonellae may be present in sludge has been shown by Findlay,[92] Hoadley (unpublished), and Hess *et al.*[93] Hoadley demonstrated approximately 10 salmonellae per 100 ml of aerobically digested municipal sludge and Hess isolated salmonellae from 91.8% of raw sludge samples, 78.1% of aerobically stabilized sludge samples, and 81.9% of anaerobically digested sludge samples. A mean of 1000 salmonellae were present per 100 ml of raw sludge, and 100 per 100 ml of digested sludge. Thus, large numbers of pathogenic bacteria, helminth ova, and viruses can be demonstrated in both primary and secondary sludges, and such sludges should not be applied to the land without disinfection or other treatment to kill pathogens.

Many studies indicate that bacteria are not removed effectively from wastewaters during primary treatment.[38,88,91] On the other hand, primary settling produces sludges that are very rich in pathogenic microorganisms.[87] Also, few viruses are removed during primary treatment.[94,95]

In general, populations of indicator bacteria are reduced by 30-40% during primary settling and by 85-99% during secondary treatment or during treatment in stabilization ponds. Sorber[96] pointed out that many pathogenic organisms survive in sewage treated by the activated sludge process, though they are greatly reduced in numbers. Indeed, a wide variety of pathogens including typhoid bacilli, cholera vibrios, tubercle bacilli, and coxsackie, polio and echoviruses have been isolated from secondary effluents. During studies of an activated sludge plant, Popp[97] demonstrated little reduction in "semiquantitative" dilution counts of salmonellae through the aeration step and secondary settling. Counts in nearly all samples of effluent from the primary settling tank, aeration chamber, and secondary settling tank contained at least 1,000 salmonellae per 100 ml, and one final effluent sample contained at least 100,000 salmonellae per 100 ml. Similarly, Pagon *et al.*[98] were unable to demonstrate the removal of salmonellae from sewage during treatment at two separate treatment plants in Switzerland from which both treated effluent and sludge were applied to the land as fertilizer. Counts of salmonellae in trickling filter and activated sludge effluents observed by Cheng *et al.*[77] varied from about 15 to 2,400 and to 4,600 organisms per 100 ml, respectively. Reductions observed by Cheng varied from about 60% to over 99%. Burns and Sierp[99] estimated that populations of shigellae may be reduced by 90 to 99% during secondary treatment. Populations of *P. aeruginosa* are reduced by about 99% during secondary treatment of sewage.[35]

Viruses appear to be removed inefficiently during biological treatment. Shuval[84] demonstrated little or no removal of enteroviruses during passage of wastewaters through a trickling filter plant and a stabilization pond, although coliform populations were reduced by about 92%. Similarly, Kott *et al.*[100] observed no reduction through stabilization ponds, and Kelly and Sanderson[101] observed little or no removal of viruses during treatment by activated sludge. Berg[102] has pointed out that it is not unusual to detect poor removals in activated sludge plants not operating properly. On the other hand, Clarke *et al.*[94] reported removal of 98% of coxsackievirus A9 and 90% of poliovirus in laboratory studies of the activated sludge process, and Mack *et al.*[95] also reported removal of viruses at times exceeding 90%.

Viruses appear to be removed effectively during chemical treatment. Removals of T4 and MS2 phages from diluted primary effluent in excess of 90% were achieved by Chaudhuri and Englebrecht[103,104] employing alum coagulation, but removals from undiluted sewage were poorer. Primavesi,[105] on the other hand, detected no virus particles in sewage after flocculation with alum and lime. Lime treatment was found by Berg *et al.*[106] to be effective in removing poliovirus 1, its effectiveness being related to the toxic effects of pH values in excess of 11. Similar observations have been made by Shelton and Drewry[107] and Lund and Ronne.[108]

It often has been suggested that disinfection of wastewaters with chlorine prior to application to soil, a process generally considered to be an effective means of controlling bacterial pathogens, is necessary to eliminate potential risks to the public health. In their studies of effluents from the Madison, Wisconsin Metropolitan Sewage Treatment Plant, Cheng *et al.*[77] failed to demonstrate salmonellae in the chlorinated effluent, although populations in the various waste streams prior to chlorination at times exceeded several thousand organisms per 100 ml. On the other hand, indicator bacteria or bacterial pathogens injured but not killed during chlorination may not recover on selective media employed for their enumeration or isolation from effluents.[109] Subsequent repair of damage to injured cells may explain in part the regrowth of coliforms and of *E. coli* in chlorinated sewage described by Shuval and his colleagues.[110] The demonstration of injury and regrowth in stored sewage suggests that chlorination may not provide reliable protection against contamination of the soil environment. This is particularly true if injured cells are as pathogenic as uninjured cells, as they appear to be. Kott[111] concluded that although disinfection of effluents from stabilization ponds with chlorine causes massive kills of bacteria, in no case are such effluents suitable for irrigation of crops eaten raw, and Sorber[96] expressed the view that chlorination, as practiced today, does not provide complete destruction of pathogenic bacteria and viruses.

Viruses may exhibit substantially greater resistance to chlorine than do the bacteria.[84,85,101,112-115] Reported concentrations of chlorine and contact times required to kill viruses in wastewaters are highly variable. Kelly and Sanderson[101] reported that resistance of virus strains varies. Thus echovirus 9 was inactivated more readily than was poliovirus,[32] and a virulent strain of poliovirus was inactivated more readily than a vaccine strain. Comparisons of research findings of different workers are complicated because concentrations of virus, content of organic matter, pH, temperature, host indicator system, and procedures for detection of virus vary from one study to another,[116] and concentrations of HOCl and chloramines may vary substantially. High concentrations of chlorine may be required to kill viruses in sewage. Shuval[117] reported that, whereas a chlorine dosage of 2 mg/l killed 99.9% of coliforms in 60 minutes, 20 mg/l were required to achieve the same kill of poliovirus. Similarly, Lindeman and Kott[118] were unable to detect any decreases in virus concentrations after 1hour in sewage treated with 8 mg of chlorine per liter.

Alternative means of killing pathogenic organisms in sewage and sludge include ozonation, gamma radiation, and pasteurization. Ozone may be a very effective virucidal agent. Shuval[117] and Berg[102] have stressed the potential usefulness of ozone based upon limited reports of research, but each calls for more work on this disinfectant. Hess and Lott[119] and Hess *et al.*[93] reported that populations of *Enterobacteriaceae* in sludge were reduced by 93-99% when heated to 70°C for 30 minutes. Limited studies of pasteurization undertaken by the Environmental Protection Agency indicate that heating of digested sludge to 75°C for 1 hour kills pathogens, and Berg[120] suggested that pasteurization was the only practical method for the control of pathogens where large amounts of solids must be penetrated, that is, in sludges. However, the use of gamma radiation has been considered, and may be useful if radioactive wastes are available as radiation sources. For example, Hess and Lott[119] employing a Cobalt 60 source, achieved 10- to 100,000-fold reductions in numbers of *Enterobacteriaceae* at 100 krad and up to 100,000,000-fold reductions at 400 to 500 krad in raw and digested sludge. Whatever methods are employed, destruction of nucleic acids of viruses probably is desirable. Melnick[121] has pointed out that viruses presumably inactivated by heat may contain infectious nucleic acids, and such virus particles may be activated within macrophages, a phenomenon analogous to the recovery of injured bacteria.

SURVIVAL OF PATHOGENS IN THE ENVIRONMENT

Because pathogenic microorganisms are present in sewage and sludges that may be applied to the land, their persistence in water, soil, sludge, feces, and aerosols is of obvious interest and concern. An understanding of the movement and persistence of pathogens will be required in order to control their dissemination among man and animals.

Aerosols may be produced where sewage or sludge is sprayed on the land or aerated. Randall and Ledbetter[122] demonstrated 1170 bacteria per cubic foot (42,000 bacteria per m^3) in samples of air obtained downwind of an activated sludge plant, in contrast to 8 organisms per cubic foot (285 organisms per m^3) in samples obtained upwind of the facility. They concluded, furthermore, that in spite of an initial "die-off," the bacteria persisted for a considerable time and distance downwind of the plant. Concentrations in the air were observed to increase with increasing wind velocity. Recently Goff *et al.*[123] published data indicating that survival times of bacteria in aerosols were greater at night than during the day. This would suggest that solar radiation causes death either by direct action on aerosolized bacteria or by action on the aerosol particles themselves. Greater aerosol emission occurred at moderate wind speeds than at high or low wind speeds. Poon[124] and Goff *et al.*[122] observed that death rates in aerosols increased with increasing temperature and at relative humidity less than 35%. Survival may be reduced also at high relative humidity, which may be attributable to breakdown of RNA.[125] Injury to the cell wall also may occur during aerosolization.[126]

Few investigations have been undertaken to explore the dissemination of microorganisms from land application sites. Merz[127] reported that bacterial travel was limited to the distance travelled by the mist from sprinklers. Sepp[128] estimated that the maximum distance travelled by bacteria was 1,000 to 1,300 ft (300 to 400 meters) in 11 mph (18 km per hr) wind and that most of the mist and bacteria settled within half that distance. Adams and Spendlove[129] were able to recover coliform organisms in the immediate vicinity of a trickling filter plant and as far as 0.8 miles (1.3 km) downwind. Of bacteria isolated 125 ft (38 m) from aeration tanks by Kenline and Scarpino,[130] 39% were *Enterobacteriaceae*, of which *Klebsiella, Enterobacter,* and *Escherichia* accounted for 78%. Of the remaining isolates, 16% were *Citrobacter* and unidentified genera, and 6% were *Shigella, Arizona, Hafnia,* and *Serratia*. No *Salmonella* or *Proteus* were isolated. Few studies have been undertaken to determine the survival of viruses in aerosols. However, Walker[131] reported that poliovirus could survive for as long as 23 hours in aerosols when the relative humidity was high.

The persistance of pathogens on vegetation and in soil has been examined more extensively. Most bacteria appear to be removed after brief passage through soil.[132-141] Pathogens can travel for miles through some limestone areas, but may be removed near the surface of heavy textured clay soils as a result of adsorption and filtration.[96] Thus the size of an organism affects its travel through soil. Many organisms probably can travel great distances through highly fractured limestone or basalt soil, but only viruses may travel through consolidated sands. Although several feet of soil appear necessary for complete removal of bacteria, 92-97% may be removed in the top 1 cm.[144] Mathur *et al.*[145] showed that 99.9% of bacteria were removed during passage of sewage through a depth of 17 inches (43 cm) and a horizontal distance of 15 inches (38 cm).

At Flushing Meadows near Phoenix, Arizona, wastewater was applied to infiltration basins lined with 3 ft (91 cm) of fine loamy sand underlain by a succession of coarse sand and gravel layers to a depth of 250 ft (76 m). Only 2% clay was present in the upper layer and the infiltration rate was 330 ft (100 m) of wastewater per year. Total coliforms decreased to a level of 2 organisms per 100 ml about 30 ft (9 m) from the point of application when basins were inundated for one day followed by a dry period of three days. Evans and Owens[147] observed a 30- to 900-fold increase in the number of fecal bacteria in subsurface pasture drainage following spray irrigation using liquid pig manure. However, populations returned to normal within two or three days after application was halted. On the other hand, Glotzbecker and Novello[148] examined leachate from a municipal landfill that contained 11 to 940 fecal coliforms and 14 to 2,200 fecal streptococci per 100 ml. The indicator bacteria persisted for hours to months in the leachate depending upon quantity, source, and temperature. Engelbrecht *et al.*,[149] using a lysimeter containing shredded municipal solid waste to simulate a sanitary landfill, demonstrated the presence of fecal coliforms and fecal streptococci in leachate for up to 96 days. However, they could not detect any virus in the leachate.

The movement of viruses through soil has not been investigated thoroughly. Results of limited percolation studies[142] suggest that complete virus removal can be achieved by careful application of sewage to land. Jopkiewicz *et al.*[143] reported that numbers of poliovirus were reduced substantially during passage through irrigation fields. Other viruses were removed less effectively. Removal of viruses in soils is a function of the characteristics of the soil. There is some indication that viruses are removed as effectively as bacteria, principally by adsorption.[134,136,137] According to Drewry and Eliassen[136] virus retention by soils is an adsorptive process that is highly efficient at pH values below 7 to 7.5, but efficiency decreases at higher pH values. Drewry and Eliassen demonstrated increased adsorption of viruses by increasing clay and silt content, ion exchange capacity, and glycerol retention capacity of soil. Merrell *et al.*[150] reported that viruses injected into percolating water were removed completely in 200 ft (61 m) of horizontal and vertical travel. Robeck *et al.*[151] observed retention of virus particles on sand beds 2 ft (61 cm) deep when the flow rate was no more than 4 linear feet (122 cm) per day. They concluded that sand filters preceeded by treatment with alum which were operated at 2 to 6 gpm per square foot (81 to 244 liters per min per m^2) retained 98-99% of virus particles. Young and Burbank[152] published an excellent study on virus removal in Hawaiian soils and showed that 95% of poliovirus II was removed during flow through 1.5 inches (3.8 cm) of Coahiawa and Lahaina soils in columns, and more than 99% of the virus particles were removed during flow through 6 inches (15 cm) of the soils, whereas only 35% of the viruses were retained by 15 inch (38 cm) columns of Tantalus soil seeded with similar quantities of the virus.

Pathogenic organisms may survive in soil and on crops for periods varying from a few hours to several months, depending upon the type of organism, soil moisture, pH, and predation and antagonism from the resident microbial flora.[153-157] Magnusson[158] reported that coliforms remained viable on grass for 15 days in dry weather and for 7 days in wet weather. On the other hand, Hess *et al.*[93] reported the survival of salmonellae on grass contaminated with sludge to be 40 to 58 weeks in a dry atmosphere. McCarty and King[159] found that enteric pathogens could survive and remain virulent for up to two months. But it has been reported also that coliforms survive longer than *S. typhi* or *M. tuberculosis,*[154,160,161] and Mallmann and Litsky[160] suggested that enterococci, rather than coliforms, appeared to be good indicators of public health hazard from sewage in soils and on vegetables. More recently, Kenner *et al.*[162] demonstrated that *E. coli* survived for at least 21 weeks after a single application of sludge to Pennsylvania fields during the spring. Furthermore, *E. coli* persisted longer than did *P. aeruginosa* (17 weeks) and salmonellae (8 weeks), and both pathogens and indicators survived longest in winter. That indicator organisms survive longer in soil during winter than during summer was demonstrated also by Van Donsel *et al.*[163]

The survivals in soil of various pathogens of man and animals, including *Erysipelothrix*,[164] *Leptospira*,[165] and *Salmonella*, have been investigated. Among these, the persistence of salmonellae has been studied most extensively. While *S. typhi* has been shown to survive for less than 24 hours in peat, and up to 2 years in frozen soils,[154] survivals of 1 month[157] and 6 to 12 months have been reported, and persistence for less than 100 days is common.[154]

Tannock and Smith,[166] investigating the survival of salmonellae, demonstrated a rapid decline in populations applied to pastures and observed a ten thousand-fold decline in the number of organisms after 10 weeks when surface water was grossly contaminated with feces, as compared to a one million-fold decrease within 2 weeks when no fecal matter was present. Salmonellae may survive for as long as 70 days in moist soil irrigated with sewage, and for 35 days in dry soil during the summer.[167] Salmonellae have been shown in numerous studies to persist for months in feces and pasture soils,[168-174] and, indeed, multiplication has been demonstrated in sewage sludge incubated at 37°C.[175]

Potential Routes of Infection

In view of the obvious opportunities for contamination of soil, groundwater, surface water, and air with pathogenic microorganisms and viruses in the vicinity of land application sites, there exists a clear need to identify possible routes of infection and to determine their significance. The contamination of surface waters ordinarily will be less than may be expected as a result of the discharge of sewage, and will not be considered here, although many of the questions raised in the following discussion may be asked about polluted surface waters as well.

The contamination of groundwaters may be a matter of considerable concern in certain areas. While pathogenic microorganisms and viruses may be removed rapidly in many soils, in limestone regions viruses may reach water supplies. The potential risks associated with contamination of groundwaters are not well defined, and outbreaks of disease related to their contamination are not well documented. Epidemiologic surveillance should be maintained in high risk areas to establish not only travel of pathogens through the ground, but rates of disease in exposed populations.

Although enteric organisms are transmitted most commonly by the oral route, it has been documented that chimpanzees can become infected with aerosolized typhoid organisms.[176] Napolitano and Rowe,[177] comparing activated sludge facilities with high rate trickling filters, concluded that more than half of the aerosol particles generated at each were more than 5μ in diameter and were not likely to enter the human lungs, but they could be deposited in the nasopharynx and swallowed. This has also been demonstrated by Sorber and Guter.[178] In view of the ability of *M. tuberculosis* to cause pulmonary tuberculosis after inhalation, Greenberg and Kupka[39] considered irrigation with sewage containing that organism to be of particular concern. Because airborne pathogens appear to remain viable and to travel for relatively short distances, the risk of exposure to airborne particles is limited primarily to workers and animals in the immediate vicinity of application sites. It is probably fair to say that, although there is little evidence of disease among workers at sewage treatment plants,[179] extensive epidemiologic studies have not been undertaken, and there exists little basis upon which to judge the true risk to workers having sporadic exposure to aerosols. Aerosols also may be a source of organisms deposited on surfaces in the vicinity of application sites. While the numbers of organisms deposited probably are small, they may be significant.

Direct contact with contaminated environments may present a risk to workers and to animals. At some heavily contaminated application sites, workers may be exposed to demonstrable risks. Hookworm and other enteric infections occur more frequently among farm workers on sewage farms in India than among the farming population in general.[180] Furthermore, Sebastian[181] noted an increased incidence of schistosomiasis in some areas of China where night soil commonly was spread on the land. Even where contamination of the environment is not as severe as in the above examples, however, organisms may spread to man and animals. That this is so is suggested by the work of Wells and James[182] who demonstrated that antibiotic resistance patterns of gut coliforms and the incidence of resistance transfer factors can be influenced greatly by contacts with animals fed antibiotics.

Although Taylor[183] concluded that the risk to the health of grazing calves of *S. dublin* in slurry applied to pasture land was not great, outbreaks of salmonellosis in cattle have been traced to fecal pollution. Schaal[184] demonstrated that the source of an outbreak of salmonellosis in a dairy herd was contaminated stream water. More recently, Bicknell[185] traced an outbreak of *S. aberdeen* in a dairy herd to pasture contaminated with overflows of sewage from a manhole. Salmonellosis in cattle associated with spreading of effluent on pastures was reported by Jack and Hepper.[186] Gibson[187] and Hughes *et al.*[188] have stressed the role of polluted water that is largely responsible for the spread of *S. dublin* in cattle, at least in South Wales. Furthermore, Findlay[92] has demonstrated salmonellae in fresh and digested sewage sludges and has suggested that risks to animals may result from use of sewage sludge as fertilizer. Risks are not limited to the animals and their handlers, however. The importance of contaminated poultry in the spread of salmonellae described earlier demonstrates that enteric pathogens may be disseminated into the general human population with meat. In a similar manner, Harvey and Price[190] observed that persistence of a *Salmonella* serotype in the sewers of an abattoir providing meat to a community was often followed by human infection.

Risks of infection among animals also are not limited to salmonellae. *Pseudomonas aeruginosa*, which may be waterborne,[192-197] can be a major cause of mastitis in dairy cows[190,191] and pneumonia in calves.[192] The application of wastes containing *P. aeruginosa* to pasture may present a risk to the health of dairy cows and calves.

The role of effluents in the infection of animals has been stressed by other workers.[198,199] Animals in contact with a reservoir of salmonellae, such as a slaughterhouse or a waste outfall, may play a role in their transmission. Salmonellae have been recovered from a high proportion of flies up to 3 miles (4.8 km) from the slaughterhouse in which they originated.[200] Nielsen [201] after studying the source of *S. typhimurium* causing substantial losses on a duck farm breeding mallards in Denmark, concluded that wild mallards, which probably included carriers, joined the pinioned stock using artificial nests and contributed contaminated eggs that were placed in incubators with the eggs of the breeding stock. They also suggested the possibility of infection of domestic animals by both sea gulls and mallards. In a later paper, Grunnett and Neilson[198] suggested that wild animals, especially sea gulls and rats that forage in sewer outfalls and dumps, transmit salmonellae from sewage to domestic animals. Muller[202] demonstrated a relationship between proximity to sewage outfalls and the frequency of carriers among gulls in Hamburg, Germany. That salmonellae might be spread similarly from land irrigated with sewage or fertilized with sludge is evident.

Furthermore, salmonellae normally can be isolated from the feces of only a small proportion of rats.[203-205] Carrier rates may be substantially higher among rats living in environments with which carrier animals are associated.[203,206] Although it was concluded by Ludlam[203] that such rats do not travel far from their homes, they may be forced to do so if their food supply is reduced. Furthermore, domestic pets may have contact with rats in their normal habitat.

The association of carriers among animals with the occurrence of salmonellae in the environment suggests that increased carrier rates may occur among dogs having access to land application sites. The potential significance of increases in carrier rates among dogs has been demonstrated by Watt and DeCapito[207] who studied endemic disease in man, which they felt probably accounted for more illness than recognized epidemics. Rectal swabs from approximately 1,300 children under 10 years of age were examined monthly. When positive cultures were obtained during the survey, domestic animals living near the patient were sampled. In more than 50% of such surveys of domestic animals, isolates were obtained of serotypes identical to those obtained from the child carrier within one block of the child's residence, suggesting a very close interrelationship between the human and animal reservoirs of salmonellae. The potential importance of dogs as potential sources of infection in man has been stressed also by Wolff *et al.*[208] Again, routes by which bacterial pathogens can be spread are suggested, and their significance should be assessed.

Crops used for human and animal consumption may become contaminated with pathogenic bacteria and viruses present in effluents and sludges. In spite of considerable reductions in the numbers of pathogens in the field and on crops as a result of exposure to ultraviolet irradiation, dessication, and competition, sufficient numbers may survive to constitute a health risk. Geldreich and Bordner[209] have reviewed sources of contamination of fruits and vegetables and provided support for establishing a limit of 1000 fecal coliforms per 100 ml in irrigation waters.

A number of outbreaks of disease attributable to the irrigation of edible crops with wastewater have been reported. Gaub[210] reported the isolation of *Shigella flexneri* from cabbage grown in fields irrigated with sewage-contaminated water. *Shigella flexneri* also was the cause of an outbreak of dysentery traced to the irrigation of pasture land with wastewater.[33] Cohen *et al.*[24] reported an outbreak of cholera El Tor in Jerusalem affecting about 250 persons. They demonstrated that vegetables grown on land irrigated with sewage were responsible for the secondary spread of the disease after the introduction of cases and carriers from outside the country. The organisms were recovered from sewage, soil irrigated with sewage, and crops from contaminated fields that were for sale in the market. Shuval[117] cited a similar outbreak of cholera in the city of Gaza. Thus, there may be a substantial risk associated with the consumption of raw vegetables grown on soil irrigated with sewage. Even crops eaten only after cooking may represent a risk since they may contaminate working surfaces and utensils in the kitchen, which in turn can lead to contamination of foods in which bacteria multiply. There appears to be no risk associated with surface irrigation of fruit trees, but if sewage is sprayed, there may be risk both to consumers and to workers.

Bovine tuberculosis can be transmitted on fodder crops irrigated with raw or partially treated wastewater.[211] However, the risk is thought to be minimal if application of wastewater is stopped 14 to 20 days prior to pasturing. There is some controversy on this point, however. Some workers have reported that the bovine tuberculosis organism can remain viable for three months in wastewater and for six months in soil. Jepsen and Roth[212] found in Denmark that *Corynebacterium bovis* infections in cattle can result from the irrigation of pasture lands with sewage. So the possibility of disease transmission to and through cattle that graze on such pasture lands should be considered.

Dunlop *et al.*[213] recovered salmonellae from about 21% of irrigation water samples contaminated with primary effluent, but not from vegetables irrigated with this water. It should, however, be noted that methods of isolation of the bacteria from water were superior to methods for vegetables. Magnusson[158] showed that from 1 to 7% of samples of vegetation were contaminated with coliforms prior to irrigation with wastewater and all were contaminated following irrigation. According to Oldham,[214] the bacteriological quality of baled hay was comparable to that of hay cut from fields irrigated with fresh water as long as the hay was not baled in a wet condition. Sorber,[96] however, considered that there was evidence of high levels of coliforms on the surfaces of vegetables irrigated with raw sewage.

In general, bacteria appear not to enter healthy and unbroken vegetables. They may, however, penetrate broken, bruised, and unhealthy plants and vegetables.[154] Once vegetables become contaminated, especially if bruised, they cannot easily be decontaminated by rinsing with water or disinfectant. Therefore, germicidal rinses with chlorine are unreliable but pasteurization at 60°C for 5 minutes is effective.[215]

Exposure to pathogens presents a risk mainly to the very young, the old, and the infirm.[3,59,73] Shooter *et al.*[216] cited circumstantial evidence that some of the *P. aeruginosa* strains colonizing hospitalized patients had been acquired from food. Similarly, Kominos *et al.*[217] isolated *P. aeruginosa* from tomatoes, radishes, celery, carrots, endive, cabbage, cucumbers, onions and lettuce obtained from the kitchen of a general hospital. These founds were implicated as the source of strains colonizing the intestinal tracts of patients. While contamination of the hands of kitchen workers, cutting boards, and knives suggested acquisition from these sources, surveillance of contamination of vegetables produced at land application sites should be maintained since contamination is possible in the field. Green *et al.*[218] demonstrated *P. aeruginosa* in 45% of soil samples from tomato fields in California where summer rainfall was negligible, but from only 1 of 24 samples from fields where other vegetables were grown. The species was isolated from only 1 of 425 samples of foliage from tomatoes growing in soil yielding *P. aeruginosa* and from 1 of 175 samples of celery growing in soil not yielding *P. aeruginosa.* Numerous other samples of vegetables failed to yield the organisms. However, while the bacteria decreased in numbers when applied to pinto beans and lettuce at low temperatures and humidities, they grew at high temperatures and humidities. Thus in hot humid climates where rainfall occurs during the growing season, splashing may cause contamination of plant parts growing above ground and growth might occur during transit. If wastewaters and sludges applied to the soil carry *P. aeruginosa*, the opportunity for contamination may be enhanced.

Viruses also may contaminate vegetables. Christovao *et al.*[219] recovered poliovirus types I and III from 5 of 11 samples of irrigation water taken from a vegetable garden in the city of Sao Paulo, Brazil and reported detection of enterovirus in soil, vegetables, and in wastewater used to irrigate the soil in which the vegetables were grown. Viruses were commonly detected on vegetables that grew close to the ground, but not on those that did not grow in contact with soil or wastewater during the growing season. Moreover, viruses may be detected more often in summer and autumn than in other seasons.[220] In extensive experiments, Mazur and Paciorkiewicz[221] detected virus on the green parts of 13 species of plants grown in soil seeded with poliovirus. The virus was isolated from 40.7% of ground parts examined and 87.9% of samples from the upper exposed parts of roots extending into water containing virus. They also found that 10% of soil samples contained virus and suggested that virus may have passed from water to soil by external capillaries along the roots.

CONCLUSIONS

Krishnaswami[211] has pointed out that "it would be presumptuous to suggest that there are no health hazards associated with the direct reuse of wastewater effluents or even that such hazards would be minimal." The reviews of Shuval,[1] Benarde,[2] Krishnaswami,[211] and Geldreich and Bordner[209] provide important background to guide attitudes toward risks to the public health associated with the application of sewage and sludge to the land. While it has been necessary to consider certain questions dealt with by these authors, it has not been our intent to repeat what they already have said. Rather, we have attempted above all to call attention to some points that appear not to have been discussed previously but which we feel provide insight into actual and potential risks to health and guidance for planning research and surveillance and for formulating regulations.

More specifically, rates of disease in the human population of the United States are low, a point frequently employed to suggest that contamination of the environment by wastes might be at an insignificant level, particularly after wastes have been treated, because our level of sanitation is high and the fecal-oral route of transmission has been interrupted. That enteric pathogens may spread easily in our population, however, has been demonstrated amply, and any change in our sanitation practices that might permit circumvention of the barriers we have erected presents a potential threat to human health. In order to minimize such risks, we must identify potential routes of transmission and populations at risk. Once we understand what these are, we can undertake the research and surveillance necessary to formulate appropriate regulations and establish controls.

Potential threats to human health associated with the pollution of surface and groundwaters, contamination of vegetables, and direct contact with aerosols and a contaminated environment have been considered elsewhere, and are discussed only briefly here. On the other hand, threats of equal concern, such as the danger to farm animals grazing on contaminated pasture, ordinarily are not considered. That disease in animals may occur as a consequence of such contamination is well established, and may result in substantial losses. Furthermore, contamination of meat in processing plants may result in the spread of enteric pathogens from infected animals to persons eating the meat. That this may occur also has been demonstrated amply. In addition, pets having access to contaminated environments may acquire enteric pathogens, carrying them into the home where they may be transmitted to small children.

Factors affecting exposure of man and animals to pathogens applied to the land with wastewater and sludge have been examined, and populations commonly occurring in sewage, their control, their survival in the environment, and numbers required to cause disease have been reviewed.

Many questions remain to be answered before sound regulations can be established. Few states at present apply regulations for the control of land application (among these are California, Arizona, Colorado, Florida, and Texas). Further research should be conducted to define more completely the ecology of many pathogens applied to land, and surveillance should be undertaken at application sites to determine the extent to which disease rates may be influenced and pathogens may be transmitted through different routes of infection. Only then can the practice be controlled effectively, and proper regulations be established.

REFERENCES

1. Shuval, H. I. "Health Factors in the Reuse of Wastewater for Agriculture, Industrial and Municipal Purposes," in *Problems in Community Wastes Management*, World Health Organization, Public Health Papers No. 38, 76 (1969).

2. Benarde, M. A. "Land Disposal of Sewage Effluent: Appraisal of Health Effects of Pathogenic Organisms," *J. Amer. Water Works Assoc.* **85**, 432 (1973).
3. Asserkoff, B., S. A. Schroeder, and P. S. Brachman. "Salmonellosis in the United States–A Five Year Review," *Amer. J. Epidemiol.* **92**, 13 (1970).
4. Committee on *Salmonella*. "An Evaluation of the *Salmonella* Problem," National Academy of Sciences, National Research Council, Washington (1969).
5. Kaufmann, A. F. and Z. L. Morrison. "An Epidemiologic Study of Salmonellosis in Turtles," *Amer. J. Epidemiol.* **84**, 364 (1966).
6. Lamm, S. H., A. Taylor, Jr., E. J. Gangarosa, H. W. Anderson, W. Young, M. H. Clark and A. R. Bruce. "Turtle Associated Salmonellosis. I. An Estimation of the Magnitude of the Problem in the United States, 1970-1971," *Amer. J. Epidemiol.* **95**, 511 (1972).
7. Altman, R., J. C. Gorman, L. L. Bernhardt, and M. Goldfield. "Turtle-Associated Salmonellosis. II. The Relationship of Pet Turtles to Salmonellosis in Children in New Jersey," *Amer. J. Epidemiol.* **95**, 518 (1972).
8. Kaufman, A. F., M. D. Fox, G. K. Morris, B. T. Wood, J. C. Feeley, and M. K. Frix. "Turtle-Associated Salmonellosis. III. The Effects of Environmental Salmonellae in Commercial Turtle Breeding Ponds," *Amer. J. Epidemiol.* **95**, 521 (1972).
9. Weissman, J. B., E. J. Gangarosa, A. Schmerler, R. L. Marier, and J. N. Lewis. "Shigellosis in Day-Care Centres," *Lancet* **1**, 88 (1975).
10. Williams, S. V., J. C. Huff, and J. A. Bryan. "Hepatitis-A and Facilities for Preschool Children," in press.
11. Jackson, C. A. W., M. J. Lindsay, and F. Shiel. "A Study of the Epizootiology and Control of *Salmonella typhimurium* Infection in a Commercial Poultry Organization," *Austr. Vet. J.* **47**, 485 (1971).
12. Clark, G. M., A. F. Kaufmann, and E. J. Gangarosa, and M. A. Thompson. "Epidemiology of an International Outbreak of *Salmonella agona*," *Lancet* **2**, 490 (1973).
13. McCoy, J. H. "Sewage Pollution of Natural Waters," in *Microbial Aspects of Pollution*, G. Sykes and F. A. Skinner, Eds. (New York: Academic Press, 1971),
14. Metzler, D., R. Culp, H. Stoltenberg, R. Woodward, G. Walton, S. Chang, N. Clarke, C. Palmer and F. Middleton. "Emergency Use of Reclaimed Water for Potable Supply at Chanute, Kan.," *J. Amer. Water Works Assoc.* **50**, 102 (1958).
15. Berger, B. B. "Public Health Aspects of Water Reuse for Potable Supply," *J. Amer. Water Works Assoc.* **52**, 599 (1960).
16. Nelson, H., H. Wilson and M. C. Robinson. "The Grazing of Cattle on Sewage Farms," *J. Proc. Inst. Sew. Purif.* **2**, 189 (1947).
17. Merrel, J. C. and P. C. Ward. "Virus Control at the Santee, California Project," *J. Amer. Water Works Assoc.* **60**, 145 (1968).
18. Reid, B. "Land Treatment and Environmental Alternatives," in *Land Disposal of Municipal Effluents and Sludges*, Proc. of a Conf. held at Rutgers University, U.S. Environmental Protection Agency Report No. 902/9-73-001 (1973).

19. Center for Disease Control. "Shigellosis Associated with Swimming in the Mississippi River–Iowa," *Weekly Morbidity and Mortality Reports* **23**, 398 (1974).
20. Barker, W. H., J. C. Sargerser, C. V. Hall, and B. J. Francis. "Food-borne Disease Surveillance, Washington State, 1969," *Amer. J. Pub. Health* **64**, 854 (1974).
21. Healy, W. A. and R. P. Grossman. "Water-Borne Typhoid Epidemic at Keene, New Hampshire," *J. New Eng. Water Works Assoc.* **75**, 37 (1961).
22. Bernard, R. P. "The Zermatt Typhoid Outbreak in 1963," *J. Hyg.* **63**, 537 (1965).
23. Barker, W. H. "Perspectives on Acute Enteric Disease: Epidemiology and Control," presented at Conference on Caribbean Epidemiological Surveillance, Jamaica (April 20-22, 1974).
24. Cohen, J., T. Schwartz, R. Klazmer, D. Pridan, H. Ghalayini, and A. M. Davies. "Epidemiological Aspects of Cholera El Tor Outbreak in a Non-Endemic Area," *Lancet* **2**, 86 (1971).
25. Wilson, W. J. and E. M. Blair. "Further Experience of the Bismuth Sulfite Media in the Isolation of *Bacillus typhosus* and *B. paratyphosus B* from Feces, Sewage and Water," *J. Hyg.* **31**, 138 (1931).
26. Green, C. E. and P. J. Beard. "Survival of *E. typhi* in Sewage Treatment Plant Processes," *Amer. J. Pub. Health* **28**, 762 (1938).
27. Kapsenberg, J. G. "Salmonellae in Treated Sewage," *Ned. Tijdschr. Geneesk* **102**, 863 (1958); *Water Poll. Abstr.* **32**, 498 (1959).
28. Coetzie, O. J. and T. Pretorius. "A Quantitative Determination of *Salmonella typhi* in Sewage and Sewage Effluents," *Pub. Health, Johannesburg* **65**, 415 (1965); *Water Poll. Abstr.* **40**, 19 (1967).
29. Moore, B. "The Detection of Paratyphoid Carriers in Towns by Means of Sewage Examination," *Monthly Bull. Ministry Health (London)* **7**, 241 (1948).
30. Harvey, R. W. S. and W. Phillips. "Survival of *Salmonella paratyphi B* in Sewers. Its Significance in the Investigation of Paratyphoid Outbreaks," *Lancet* **2**, 137 (1955).
31. Buss, W. and T. Inal. "Testing the Efficiency of a Municipal Sewage Works by Examining the Sewage for *Salmonella* as a Form of Stage Control," *Berl. Munch Tierarztl. Wschr.* **70**, 311 (1957), *Water Poll. Abstr.* **31**, 1609 (1958).
32. Kabler, P. W. "Removal of Pathogenic Microorganisms by Sewage Treatment Processes," *Sew. Ind. Wastes* **31**, 1373 (1959).
33. Browning, G. E. and J. O. Mankin. "Gastroenteritis Epidemic Owing to Sewage Contamination of Public Water Supply," *J. Amer. Water Works Assoc.* **58**, 1465 (1966).
34. Northington, C. W., S. L. Chang and L. T. McCabe. "Health Aspects of Wastewater Reuse," in *Water Quality Improvement by Physical and Chemical Processes,* E. F. Gloyna and W. W. Eckenfelder, Eds. (University of Texas Press, 1970).
35. Hoadley, A. W., E. McCoy, G. A. Rohlich. "Untersuchungen uber *Pseudomonas aeruginosa* in Oberflachengewassern. I. Quellen," *Arch. Hyg. Bakteriol.* **152**, 238 (1968).

36. Rhiner, C. "The Longevity of Tubercle Bacilli in Sewage and Stream Water," *Amer. Rev. Tuberc.* **31**, 493 (1935).
37. Jensen, K. A. and K. E. Jensen. "Occurrence of Tubercle Bacilli in Sewage and Experiments on Sterilization of Tubercle Bacilli-Containing Sewage with Chlorine," *Acta Tuberc. Scand.* **16**, 217 (1942).
38. Kelly, S. M., M. E. Clark, and M. B. Coleman. "Demonstration of Infectious Agents in Sewage," *Amer. J. Pub. Health* **45**, 1438 (1955).
39. Greenberg, A. E. and E. Kupka. "Tuberculosis Transmission by Wastewaters–A Review," *Sew. Ind. Wastes* **29**, 524 (1957).
40. Cleere, R. L. "Resume of Swimming Pool Granuloma," *Sanitation* **23**, 105 (1960).
41. Bhaskaran, T. R., M. N. Lahiri, and B. K. G. Roy. "Effect of Sewage Treatment Processes on Survival of Tubercle Bacilli," *Indian J. Med. Res.* **48**, 790 (1960); *Water Poll. Abstr.* **34**, 2097 (1961).
42. Coin, L., M. L. Menetrier, J. Labonde, and M. C. Hannoun. "Modern Microbiological and Virological Aspects of Water Pollution," *Proc. 2nd Internat. Conf. Water Poll. Res.*, Tokyo, **1**, 1 (1965).
43. Bryan, F. L. "What the Sanitarian Should Know about *Clostridium perfringens* Foodborne Illness," *J. Milk Food Technol.* **32**, 381 (1969).
44. Bonde, G. "Bacterial Indicators of Water Pollution," *Teknisk Forlag,* Copenhagen (1963).
45. Nussbaumer, N. L. "Building for 16 m.g.d., Designing for 24 m.g.d., Planning for 32 m.g.d.," *Water Works Eng.* **116**, 722 (1963), *Water Poll. Abstr.* **38**, 18 (1965).
46. Grabow, W. O. K. "The Virology of Waste Water Treatment," *Water Res.* **2**, 675 (1968).
47. Berg, G. "Reassessment of the Virus Problem in Sewage and in Surface and Renovated Waters," in *Water Quality: Management and Pollution Control Problems,* S. H. Jenkins, Ed. (Oxford: Pergamon Press, 1973).
48. Wang, W. L. and S. G. Dunlop. "Animal Parasites in Sewage and Irrigation Water," *Sew. Ind. Wastes* **26**, 1020 (1954).
49. Dunlop, S. G. and W. L. Wang. "Studies on the Use of Sewage Effluent for Irrigation of Truck Crops," *J. Milk Food Technol.* **24**, 44 (1961); *Water Poll. Abstr.* **35**, 1372 (1962).
50. Kott, H. and Y. Kott. "Detection and Viability of *Endamoeba histolytica* Cysts in Sewage Effluents," *Water Sew. Works* **114**, 177 (1967).
51. Rowan, W. B. "Sewage Treatment and Schistosome Eggs," *Amer. J. Trop. Med. Hyg.* **13**, 572 (1964).
52. Stringer, R. and C. W. Kruse. "Amoebic Cysticidal Properties of Halogens in Water," *Proc. National Specialty Conf. on Disinfection,* (New York: American Society of Civil Engineers, 1970), p. 319.
53. Cram. E. B. "The Effect of Various Treatment Processes on the Survival of Helminth Ova and Protozoan Cysts in Sewage," *Sew. Works J.* **15**, 1119 (1943).
54. Chang, S. L. "Viruses, Amoebas, and Nematodes and Public Water Supplies," *J. Amer. Water Works Assoc.* **53**, 288 (1961).
55. Chang, S. L. and P. W. Kabler. "Free-Living Nematodes in Aerobic Treatment Plant Effluent," *J. Water Poll. Control Fed.* **34**, 1256 (1962).

56. Liebman, H. "Parasites in Sewage and the Possibilities of Their Extinction," *Adv. in Water Poll. Res.* **2**, 269 (1965).
57. Murad, J. and G. Bazer. "Diplogasterid and Rhabditid Nematodes in Wastewater Treatment Plant and Factors Related to their Dispersal," *J. Water Poll. Control Fed.* **42**, 106 (1970).
58. Sladka, A. and V. Ottova. "The Most Common Fungi in Biological Treatment Plants," *Hydrobiologia* **31**, 350 (1968).
59. Reller, L. B., E. J. Gangarosa, and P. S. Brachman. "Shigellosis in the United States: Five-Year Review of Nationwide Surveillance, 1964-1968," *Amer. J. Epidemiol.* **91**, 161 (1970).
60. Levine, M. M., H. L. DuPont, M. Khodabandelou, R. B. Hornick. "Long-Term Shigella-Carrier State," *New Eng. J. Med.* **288**, 1169 (1973).
61. Craun, G. F. and L. J. McCabe. "Review of the Causes of Water-borne-Disease Outbreaks," *J. Amer. Water Works Assoc.* **65**, 74 (1973).
62. Felsenfeld, O. "A Review of Recent Trends in Cholera Research and Control," *Bull. World Health Org.* **34**, 161 (1966).
63. Pal, S. C., B. S. Misra, D. D. Arora, K. V. Arora, S. Pattanayak, C. G. Pandit and J. B. Shrivastav. "Studies on Cholera Carriers in Delhi," *Indian J. Med. Res.* **61**, 1 (1973).
64. Jukes, T. H. "The Present Status and Background of Antibiotics in the Feeding of Domestic Animals," in "The Problem of Drug-Resistant Pathogenic Bacteria," E. L. Dulaney and A. I. Laskin, Eds. *Ann. N.Y. Acad. Sci.* **182**, 362 (1971).
65. Jukes, T. H. "Antibiotics in Animal Feeds and Animal Production," *Bioscience* **22**, 526 (1972).
66. Anderson, E. S. "The Ecology of Transferable Drug Resistance in the Enterobacteria," *Ann. Rev. Microbiol.* **22**, 131 (1968).
67. Walton, J. R. "The Public Health Implications of Drug-Resistant Bacteria in Farm Animals," in "The Problem of Drug-Resistant Pathogenic Bacteria," E. L. Dulaney and A. I. Laskin, Eds. *Ann. N.Y. Acad. Sci.* **182**, 358 (1971).
68. Grabow, W. O. K., O. W. Prozesky, and L. S. Smith. "Review Paper: Drug Resistant Coliforms Call for Review of Water Quality Standards," *Water Res.* **8**, 1 (1974).
69. Sturtevant, A. B., Jr. and T. W. Feary. "Incidence of Infectious Drug Resistance Among Lactose-Fermenting Bacteria Isolated from Raw and Treated Sewage," *Appl. Microbiol.* **18**, 918 (1969).
70. Sturtevant, A. B., Jr.,G. H. Cassell and T. W. Feary. "Incidence of Infectious Drug Resistance Among Fecal Coliforms Isolated from Raw Sewage," *Appl. Microbiol.* **21**, 487 (1971).
71. Grabow, W. O. K. and O. W. Prozesky. "Drug Resistance of Coliform Bacteria in Hospital and City Sewage," *Antimicrob. Agents Chemotherapy* **3**, 175 (1973).
72. DuPont, H. L. and R. B. Hornick. "Clinical Approach to Infectious Diarrheas," *Medicine* **52**, 265 (1973).
73. Lang, D. J., L. S. Kunz, A. R. Martin, S. A. Schroeder, and A. Thompson. "Carmine as a Source of Nosocomial Salmonellosis," *New Eng. J. Med.* **276**, 829 (1967).

74. Hornick, R. B., S. E. Geisman, T. E. Woodward, H. L. DuPont, A. T. Dawkins, and M. J. Snyder. "Typhoid Fever: Pathogenesis and Immunologic Control," *New Eng. J. Med.* **283**, 686 (1970).
75. Hoadley, A. W. "*Pseudomonas aeruginosa* in Surface Waters," in press.
76. Kehr, R. W. and C. T. Butterfield. "Notes on the Relation Between Coliforms and Enteric Pathogens," *U.S. Pub. Health Rep.* **58**, 589 (1943).
77. Cheng, C. M., W. C. Boyle, and J. M. Goepfert. "Rapid Quantitative Method for *Salmonella* Detection in Polluted Waters," *Appl. Microbiol.* **21**, 622 (1971).
78. Center for Disease Control. "Reported Morbidity and Mortality in the United States 1973," *Weekly Morbidity and Mortality Reports* **22**, 53 (1974).
79. World Health Organization. "Reuse of Effluents: Methods of Wastewater Treatment and Health Safeguards–Report of a W.H.O. Meeting of Experts," *Tech. Rep. Ser. No. 517*, Geneva, Switzerland (1973).
80. Maxcy, K. F. and H. A. Howe. "The Significance of the Finding of the Virus of Infantile Paralysis in Sewage: A Review," *Sew. Works J.* **15**, 1101 (1943).
81. Kollins, S. A. "The Presence of Human Enteric Viruses in Sewage and Their Removal by Conventional Sewage Treatment Methods," *Adv. Appl. Microbiol.* **8**, 145 (1966).
82. Kelly, S. and W. W. Sanderson. "Density of Enteroviruses in Sewage," *J. Water Poll. Control Fed.* **32**, 1269 (1960).
83. Wolf, H. W. "Biological Aspects of Water," *J. Amer. Water Works Assoc.* **63**, 181 (1971).
84. Shuval, H. I. "Detection and Control of Entero-Viruses in the Water Environment," in *Developments in Water Quality Research*, H. I. Shuval, Ed., Proc. Intl. Conf. on Water Quality and Pollution Research, Jerusalem (Ann Arbor, Mich.: Ann Arbor Science Publishers, 1970).
85. Clarke, N. A. and P. W. Kabler. "Human Enteric Viruses in Sewage," *Health Lab. Sci.* **1**, 44 (1964).
86. American Water Works Association Committee Report. "Viruses in Water," *J. Amer. Water Works Assoc.* **61**, 491 (1969).
87. Pramer, D., H. Heukelekian, and R. A. Ragozkie. "Survival of Tubercle Bacilli in Various Sewage Treatment Processes, I. Development of a Method for the Quantitative Recovery of *Mycobacteria* from Sewage," *U.S. Pub. Health Rep.* **65**, 851 (1950).
88. Bloom, H. H., W. N. Mack, and W. L. Mallmann. "Enteric Viruses and *Salmonellae* Isolations, II. Media Comparison for *Salmonellae,*" *Sew. Ind. Wastes* **30**, 1455 (1958).
89. Lung, E. "Observations on the Virus Binding Capacity of Sludge," in *Advances in Water Pollution Research*, S. H. Jenkins, Ed. Proc. of the 5th Internat. Conf. held at San Francisco and Hawaii in 1970 (Oxford: Pergamon Press, 1971).
90. Manwaring, J. F., M. Chauduri, and R. S. Engelbrecht. "Removal of Virus by Coagulation and Flocculation," *J. Amer. Water Works Assoc.* **63**, 298 (1971).

91. Heukelekian, H. and M. Albanese. "Enumeration and Survival of Human Tubercle Bacilli in Polluted Waters, II. Effect of Sewage Treatment and Natural Purification," *Sew. Ind. Wastes* **28**, 1094 (1956).
92. Findlay, C. R. "Salmonellae in Sewage Sludge: Part I, Occurrence," *Vet. Rec.* **93**, 100 (1973).
93. Hess, E., G. Lott, and C. Breer. "Klarschlamm und Freilandbiologie von Salmonellen," *Zentralbl Bakteriol. Hyg., I Abt. Orig. B* **158**, 446 (1974).
94. Clarke, N. A., R. E. Stevenson, S. L. Chang, and P. W. Kabler. "Removal of Enteric Viruses from Sewage by Activated Sludge Treatment," *Amer. J. Pub. Health* **58**, 1118 (1961).
95. Mack. W. N., J. R. Frey, B. J. Riegle, and W. L. Mallmann. "Enterovirus Removal by Activated Sludge Treatment," *J. Water Poll. Control Fed.* **34**, 1133 (1962).
96. Sorber, C. A. "Protection of the Public Health," in *Land Disposal of Municipal Effluents and Sludges,* Proc. of a Symposium held at Rutgers Univ., U.S. Environmental Protection Agency Report No. 902/9-73-001 (1973).
97. Popp, L. "Uber die Elimination von Salmonellen durch biologische Abwasserbehandlung," *Zentralbl. Bakteriol. Hyg., I Abt. Orig. B* **157**, 184 (1973).
98. Pagon, S., W. Sonnabend, and U. Krech. "Epidemiologische Zusammenhänge Zwischen Men Schlichen und Tierischen Salmonella-Ausscheidern und Deren Umwelt im Schweizerischen Bodenseeraum," *Zentralbl. Bakteriol. Hyg. I Abt. Orig. B* **158**, 395 (1974).
99. Burns, H. and F. Sierp. "Influence of the Activated Sludge Process on Pathogenic Bacteria," *Zeit. Hyg. Infekt.-Krankh* **107**, 4 (1927), *Water Poll. Abstr.* **1**, B-33 (1928).
100. Kott, Y., N. Roze, S. Sperber, and N. Betzer. "Bacteriophages as Viral Pollution Indicators," *Water Res.* **8**, 165 (1974).
101. Kelly, S. and W. W. Sanderson. "The Effect of Sewage Treatment on Viruses," *Sew. Ind. Wastes* **31**, 683 (1959).
102. Berg, G. "Removal of Viruses from Sewage, Effluents, and Waters. 1. A Review," *Bull. World Health Org.* **49**, 451 (1973).
103. Chaudhuri, M. and R. S. Engelbrecht. "Removal of Virus from by Chemical Coagulation and Flocculation," *J. Amer. Water Works Assoc.* **62**, 563 (1970).
104. Chaudhuri, M. and R. S. Engelbrecht. "Virus Removal in Wastewater Renovation by Chemical Coagulation and Flocculation," in *Advances in Water Pollution Research*, S. H. Jenkins, Ed., Proc. of the 5th Internat. Conf. held in San Francisco and Hawaii in 1970 (Oxford: Pergamon Press, 1971).
105. Primavesi, C. A. "Destruction of Virus in River Water During Flocculation and Chlorination," *Gesundheits-Ingenieur* **91**, 266 (1970).
106. Berg, G., R. G. Dean, and D. R. Dahling. "Removal of Poliomyelitis I from Secondary Effluent by Lime Flocculation and Rapid Sand Filters," *J. Amer. Water Works Assoc.* **60**, 193 (1968).
107. Shelton, S. P. and W. A. Drewry. "Tests of Coagulants for the Reduction of Viruses, Turbidity, and Chemical Oxygen Demand," *J. Amer. Water Works Assoc.* **65**, 627 (1973).

108. Lund, E. and V. Ronne. "On the Isolation of Virus from Sewage Treatment Plant Sludges," *Water Res.* 7, 863 (1973).
109. Braswell, J. R. and A. W. Hoadley. "Recovery of *Escherichia coli* from Chlorinated Secondary Sewage," *Appl. Microbiol.* **28**, 328 (1974).
110. Shuval, H. I., J. Cohen, and R. Kolodney. "Regrowth of Coliforms and Fecal Coliforms in Chlorinated Wastewater Effluent," *Water Res.* 7, 537 (1973).
111. Kott, Y. "Chlorination of Sewage Oxidation Pond Effluent," in *Development in Water Quality Research*, H. I. Shuval, Ed., Proc. Intl. Conf. on Water Quality and Pollution Research, Jerusalem (Ann Arbor, Mich.: Ann Arbor Science Publishers, 1970).
112. Weidenkopf, S. J. "Inactivation of Type I Poliomyelitis Virus with Chlorine," *Virology* **5**, 56 (1958).
113. Morris, J. C. "Future of Chlorination," *J. Amer. Water Works Assoc.* **58**, 1475 (1966).
114. Dunlop, S. G. "Survival of Pathogens and Related Disease Hazards," *Proc. of a Symposium on Municipal Sewage Effluents for Irrigation* Louisiana Polytechnic Institute, Ruston (1969).
115. Durham, D. and H. W. Wolf. "Wastewater Chlorination: Panacea or Placebo?" *Water Sew. Works* **120**, 67 (1973).
116. Varma, M. M., B. A. Christian, and D. W. McKinstry. "Inactivation of Sabin Oral Polio Type I Virus," *J. Water Poll. Control Fed.* **46**, 987 (1974).
117. Shuval, H. I. "Disinfection of Wastewater for Agricultural Utilization," (1974).
118. Lindeman, S. and Y. Kott. "The Effect of Chlorination on Enteroviruses in the Effluents of the Haifa Sewage Treatment Plant," *Israel J. Med. Sci.* 7, 111 (1971).
119. Hess, E. and G. Lott. "Klärschlamm aus der Sicht des Veterinärhygienikers," *Gas Wasser Abwasser* **51**, 62 (1971).
120. Berg, G. "Removal of Viruses from Sewage, Effluents, and Waters. 2. Present and Future Trends," *Bull. World Health Org.* **49**, 461 (1973).
121. Melnick, J. L. Discussion following paper by G. Berg entitled "Reassessment of the Virus Problem in Sewage and in Surface and Renovated Waters," in *Water Quality: Management and Pollution Control Problems,* S. H. Jenkins, Ed. (Oxford: Pergamon Press, 1973).
122. Randall, C. W. and J. O. Ledbetter. "Bacterial Air Pollution from Activated Sludge Unit," *J. Amer. Ind. Hyg. Assoc.* **27**, 506 (1966).
123. Goff, G. D., J. C. Spendlove, A. P. Adams and P. S. Nicholes. "Emission of Microbial Aerosols from Sewage Treatment Plants that Use Trickling Filters," *Health Serv. Rep.* **88**, 640 (1973).
124. Poon, C. P. C. "Viability of Long Storaged Airborne Bacterial Aerosols," *J. San. Eng. Div., Amer. Soc. Civil Engr.* **94**, 1137 (1968).
125. Cox, C. S. "The Cause of Loss of Viability of Airborne *Escherichia coli* K12," *J. Gen. Microbiol.* **57**, 77 (1969).
126. Hambleton, P. "Repair of Wall Damage in *Escherichia coli* Recovered from an Aerosol," *J. Gen. Microbiol.* **69**, 81 (1971).
127. Merz, R. C. "Third Report on the Study of Wastewater Reclamation and Utilization," California State Water Poll. Control Board Publication No. 18, Sacramento, California (1957).

128. Sepp, E. "The Use of Sewage for Irrigation–A Literature Review," Bureau of Sanitary Engineering, California State Dept. of Public Health, Sacramento, California (1971).
129. Adams, A. P. and J. C. Spendlove. "Coliform Aerosols Emitted by Sewage Treatment Plants," *Science* **169**, 1218 (1970).
130. Kenline, P. A. and P. V. Scarpino. "Bacterial Air Pollution from Sewage Treatment Plants," *J. Amer. Ind. Hyg. Assoc.* **33**, 346 (1972).
131. Walker, B. "Viruses Respond to Environmental Exposure," *J. Environ. Health.* **32**, 532 (1970).
132. Krone, R. B., P. H. McGauhey, and H. B. Gotaas. "Direct Discharge of Groundwater with Sewage Effluent," *J. San. Eng. Div., Amer. Soc. Civil Eng.* **83**, 1 (1957).
133. Krone, R. B., G. T. Orlob, and C. Hodkinson. "Movement of Coliform Bacteria Through Porous Media," *Sew. Ind. Wastes* **30**, 1 (1958).
134. Eliassen, R., P. Kruger and W. Drewry. "Studies on the Movement of Viruses in Groundwater," Progress Report to the Commission on Environmental Hygiene of the Armed Forces Epidemiological Board (1965).
135. McMichael, F. C. and J. E. McKee. "Wastewater Reclamation at Whittier Narrows," California State Water Quality Control Board Publication No. 33 (1966).
136. Drewry, W. A. and R. Eliassen. "Virus Movement in Ground Water," *J. Water Poll. Control Fed.* **40**, R257 (1968).
137. Krone, R. B. "The Movement of Disease Producing Organisms Through Soils," *Proc. of a Symposium on Municipal Sewage Effluent for Irrigation,* Louisiana Polytechnic Institute, Ruston (1969).
138. Romero, J. C. "The Movement of Bacteria and Viruses Through Porous Media," *Ground Water* **8**, 37 (1970).
139. Wesner, G. M. and D. C. Baier. "Injection of Reclaimed Wastewater into Confined Aquifers," *J. Amer. Water Works Assoc.* **62**, 203 (1970).
140. Stevens, R. M. "Green Land: Clean Streams. A Report," Center for the Study of Federalism, Temple University, Philadelphia (1972).
141. Bouwer, H. "Renovating Secondary Effluent by Groundwater Recharge with Infiltration Basins," in *Recycling Treated Municipal Wastewater and Sludge Through Forest and Cropland*, W. E. Sopper and L. T. Kardos, Eds. (University Park, Pa.: Pennsylvania State University Press, 1973).
142. Water Resources Task Group. "A Technical Evaluation of Land Disposal of Wastewater and the Needs for Planning and Monitoring Water Resources in Dane County, Wisconsin," The Dane County Regional Planning Commission, Madison, Wisconsin (1971).
143. Jopkiewicz, T., K. Krzeminska and Z. Stachowska. "A Virological Survey of Sewage in the City of Bydgoszcz," *Przegl. Epidem.* **22**, 521 (1968); *Excerpta Medica* **15**(17), 3477 (1969).
144. Marculeseu, I. and A. Drucan. "Investigations Using Labelled Bacteria in the Study of Irrigation with Sewage," *Stud. Prot. Epurarea Apelor,* Bucharest, **59** (1962).
145. Mathur, R. P., S. Chandra, and K. A. Bhardwaj. "Two Dimensional Study of Travel of Pollutants in Roorkee Soil," *J. Institution Engr.* (India) **48**, 3 (1968).

146. Bouwer, H. "Putting Wastewater to Beneficial Use—The Flushing Meadows Project," *Proc. 12th Arizona Watershed Symposium,* 25 (1968).
147. Evans, M. R. and J. D. Owens. "Factors Affecting the Concentration of Fecal Bacteria in Land Drainage Water," *J. Gen. Microbiol.* **71**, 477 (1972).
148. Glotzbecker, R. A. and A. L. Novello. "Poliovirus and Bacterial Indicators of Fecal Pollution in Landfill Leachates," *News of Environmental Research in Cincinnati*, Technical Information Office, National Environmental Research Center, U.S.E.P.A., Cincinnati, Ohio (Jan. 31, 1975).
149. Engelbrecht, R. S., M. S. Weber, P. Amirhor, D. H. Foster, and D. LaRossa. "Biological Properties of Sanitary Landfill Leachates," presented at Water Resources Symposium No. 7: Virus Survival in Water and Wastewater Systems, University of Texas, Austin April 1-3, 1974.
150. Merrell, J. C., W. F. Jopling, R. F. Bott, A. Katko, and H. E. Pintler. "The Santee Recreation Project, Santee, California. Final Report," U.S. Dept. Interior, Water Poll. Control Res. Publ. No. WP-20-7 (1967).
151. Robeck, G. G., H. R. Clark, and K. A. Dostal. "Effectiveness of Water Treatment Processes in Virus Removal," *J. Amer. Water Works Assoc.* **54**, 1275 (1962).
152. Young, R. H. F. and N. C. Burbank, Jr. "Virus Removal in Hawaiian Soils," *J. Amer. Water Works Assoc.* **65**, 598 (1973).
153. Falk, L. L. "Bacterial Contamination of Tomatoes Grown in Polluted Soil," *Amer. J. Pub. Health* **39**, 1338 (1949).
154. Rudolfs, W., L. L. Falk and R. A. Ragotzkie. "Literature Review on the Occurrence and Survival of Enteric, Pathogenic, and Relative Organisms in Soil, Water, Sewage, and Sludges, and on Vegetation. I, Bacterial and Virus Diseases," *Sew. Ind. Wastes* **22**, 1261 (1950).
155. Rudolfs, W., L. L. Falk and R. A. Ragotzkie. "Contamination of Vegetables Grown in Polluted Soil. I. Bacterial Contamination," *Sew. Ind. Wastes* **23**, 253 (1951).
156. Dunlop, S. G. "The Irrigation of Truck Crops with Sewage Contaminated Water," *The Sanitarian* **15,** 107 (1952).
157. McGauhey, P. H. and R. B. Krone. "Report of Investigation of Travel of Pollution," California State Water Pollution Control Board Publication No. 11 (1954).
158. Magnusson, F. "Spray Irrigation of Dairy Effluent," *Ann. Bull. Internat. Dairy Fed.* **77**, 122 (1974).
159. McCarty, P. L. and P. H. King. "The Movement of Pesticides in Soils," *Proc. 21st Ind. Waste Conference,* (Lafayette, Ind.: Purdue University, 1966), p. 156.
160. Mallmann, W. L. and W. Litsky. "Survival of Selected Enteric Organisms in Various Types of Soil," *Amer. J. Publ. Health* **41**, 38 (1951).
161. Mallmann, W. L. and W. N. Mack. "Biological Contamination of Ground Water," in *Groundwater Contamination,* U.S. Public Health Services Technical Report No. W61-5 (1961).
162. Kenner, B. A., G. K. Dotson and J. E. Smith. "Simultaneous Quantitation of *Salmonella* Species and *Pseudomonas aeruginosa*," U.S.

Environmental Protection Agency, National Environmental Research Center, Cincinnati, Ohio (1971).
163. Van Donsel, D. J., E. E. Geldreich, and N. A. Clarke. "Seasonal Variations in Survival of Indicator Bacteria in Soil and their Contribution to Stormwater Pollution," *Appl. Microbiol.* **15**, 1362 (1967).
164. Wood, R. L. "Survival of *Erysipelothrix rhusiopatiae* in Soil under Various Environmental Conditions," *Cornell Vet.* **63**, 390 (1973).
165. Diesch, S. L. "Survival of Leptospires in Cattle Manure," *J. Amer. Vet. Med. Assoc.* **159**, 1513 (1971).
166. Tannock, G. W. and J. M. B. Smith. "Studies on the Survival of *Salmonella typhimurium* and *Salmonella bovismorbificans* on Pasture and in Water," *Aust. Vet. J.* **47**, 557 (1971).
167. Bergner-Rabinowitz, S. "The Survival of Coliforms, *Streptococcus faecalis* and *Salmonella tennessee* in the Soil and Climate of Israel," *Appl. Microbiol.* **4**, 101 (1956).
168. Mair, N. S. and A. I. Ross. "Survival of *Salm. typhimurium* in the Soil," *Monthly Bull. Ministry Health (London)* **19**, 39 (1960).
169. Thomas, K. L. "Survival of *Salmonella paratyphi B*, Phage Type 1 var. 6, in Soil," *Monthly Bull. Ministry Health (London)* **26**, 39 (1967).
170. Findlay, C. R. "The Survival of *Salmonella dublin* in Cattle Slurry," *Vet. Rec.* **89**,224 (1971).
171. Taylor, R. J. and M. R. Burrows. "The Survival of *Escherichia coli* and *Salmonella dublin* in Slurry on Pasture and the Infectivity on *S. dublin* for Grazing Calves," *Brit. Vet. J.* **127**, 536 (1971).
172. Jeffrey, D. C. "Persistence of Three *Salmonella* spp. in Bovine Faeces," *Vet. Rec.* **88**, 329 (1971).
173. Findlay, C. R. "The Persistence of *Salmonella dublin* in Slurry in Tanks and on Pasture," *Vet. Rec.* **91**, 233 (1972).
174. Tannock, G. W. and J. M. B. Smith. "Studies on the Survival of *Salmonella typhimurium* and *Salmonella bovis-morbificans* on Soil and Sheep Faeces," *Res. Vet. Sci.* **13**, 150 (1972).
175. Findlay, C. R. "Salmonella in Sewage Sludge: Part II. Multiplication," *Vet. Rec.* **93**, 102 (1973).
176. Crozier, D. and T. E. Woodward. "Activities of the Commission on Epidemiological Survey, 1961," *Military Med.* **127**, 701 (1962).
177. Napolitano, P. J. and D. R. Rowe. "Microbial Content of Air Near Sewage Treatment Plants," *Water Sew. Works* **113**, 480 (1966).
178. Sorber, C. A. and K. J. Guter. "Health and Hygiene Aspects of Spray Irrigation," in *Wastewater Management by Disposal on the Land*, S. C. Reed, Coordinator, Cold Regions Res. and Eng. Lab (Hanover, N.H.: U.S. Army Corps of Engineers, 1972).
179. Dixon, F. R. and L. J. McCabe. "Health Aspects of Wastewater Treatment," *J. Water Poll. Control Fed.* **36**, 984 (1964).
180. Central Public Health Engineering Research Institute. "Health Status of Sewage Farm Workers," *Technical Digest* No. 17, Nagpur, India (1971).
181. Sebastian, F. P. "Modern Technology Battles Ancient Traditions," *Water Wastes Eng.* **10**, 20 (1973).
182. Wells, D. M. and O. B. James. "Transmission of Infectious Drug Resistance from Animals to Man," *J. Hyg.* **71**, 209 (1973).

183. Taylor, R. J. "A Further Assessment of the Potential Hazard for Calves Allowed to Graze Pasture Contaminated with *Salmonella dublin* in Slurry," *Brit. Vet. J.* **129**, 354 (1973).
184. Schaal, E. "Uber eine durch Backwasser verursachte Salmonella-Enzootie in einem Rinderbestand," *Deutsche Tierärztliche Wochenschrift* **70**, 267 (1963).
185. Bicknell, S. R. "*Salmonella aberdeen* Infection in Cattle Associated with Human Sewage," *J. Hyg.* **70**, 121 (1972).
186. Jack, E. J. and P. T. Hepper. "An Outbreak of *Salmonella typhimurium* Infection in Cattle Associated with Spreading of Slurry," *Vet. Rec.* **84**, 196 (1969).
187. Gibson, E. A. "Salmonellosis in Calves," *Vet. Rec.* **73**, 1284 (1961).
188. Hughes, L. E., E. A. Gibson, H. E. Roberts, E. T. Davies, G. Davies, and W. J. Sojka. "Bovine Salmonellosis in England and Wales," *Brit. Vet. J.* **127**, 225 (1971).
189. Harvey, R. W. S. and T. H. Price. "Sewer and Drain Swabbing as a Means of Investigating Salmonellosis," *J. Hyg.* **68**, 611 (1970).
190. Howell, D. "Survey on Mastitis Caused by Environmental Bacteria," *Vet. Rec.* **90**, 654 (1972).
191. Szazados, I. and I. Kadas. "Role of *Pseudomonas aeruginosa* in Actinomycosis-like Bovine Mastitis," *Acta Vet. Acad. Scient. Hungar.* **22**, 241 (1972).
192. Prasad, B. M., C. P. Srivastava, K. G. Narayan, and A. K. Prasad. "Source of *Pseudomonas* Infection in Calves," *Indian J. Anim. Health* **7**, 51 (1968).
193. Pickens, E. M., M. F. Welsh, and L. J. Poelma. "Pyocyaneus Bacillosis and Mastitis due to *Ps. aeruginosa,*" *Cornell Vet.* **16**, 186 (1926).
194. Cherrington, V. A. and E. M. Gildow. "Bovine Mastitis Caused by *Pseudomonas aeruginosa,*" *J. Amer. Vet. Med. Assoc.* **79**, 803 (1931).
195. Hoadley, A. W. and E. McCoy. "Some Observations on the Ecology of *Pseudomonas aeruginosa* and its Occurrence in the Intestinal Tracts of Animals," *Cornell Vet.* **58**, 354 (1968).
196. Curtis, P. E. "*Pseudomonas aeruginosa* Contamination of Warm Water System Used for Pre-Milking Udder Washing," *Vet. Rec.* **84**, 476 (1969).
197. Malmo, J. "An Outbreak of Mastitis due to *Pseudomonas aeruginosa* in a Dairy Herd," *Amer. Vet. J.* **48**, 137 (1972).
198. Grunnet, K. and B. Brest Nielsen. "*Salmonella* Types Isolated from the Gulf of Aarhus Compared with Types from Infected Human Beings, Animals, and Feed Products in Denmark," *Appl. Microbiol.* **18**, 985 (1969).
199. Edel, W., A. M. Guinee, M. Van Schothorst, and E. H. Kampelmacher. "The Role of Effluents in the Spread of Salmonellae," *Zentralbl. Bakteriol. Hyg. I Abt. Orig.* **221**, 547 (1972).
200. Greenberg, B. and V. Miggiano. "Host-Contaminant Biology of Muscoid Flies. IV. Microbial Competition in a Blowfly," *J. Infect. Dis.* **112**, 37 (1963).
201. Nielsen, B. Brest. "*Salmonella typhyimurium* Carriers in Seagulls and Mallards as a Possible Source of Infection of Domestic Animals," *Nord. Vet.-Med.* **12**, 417 (1960).

202. Muller, G. "Salmonella in Bird Faeces," *Nature* **207**, 1315 (1965).
203. Ludlam, G. B. "*Salmonella* in Rats, with Special Reference to Findings in a Butcher's By-Products Factory," *Monthly Bull. Ministry Health (*London) **13**, 196 (1954).
204. Lee, P. E. "*Salmonella* Infections of Urban Rats in Brisbane, Queensland," *Aust. J. Exp. Biol.* **33**, 113 (1955).
205. Schnurrenberger, P. R., L. J. Held, R. J. Martin, K. D. Quist, and M. M. Galton. "Prevalence of *Salmonella* spp. in Domestic Animals and Wildlife on Selected Illinois Farms," *J. Amer. Vet. Med. Assoc.* **153**, 422 (1968).
206. Goyal, S. M. and I. P. Singh. "Probable Sources of Salmonellae on a Poultry Farm," *Brit. Vet. J.* **126**, 180 (1970).
207. Watt, J. and T. DeCapito. "The Frequency and Distribution of *Salmonella* Types Isolated from Man and Animals in Hidalgo County, Texas," *Amer. J. Hyg.* **51**, 343 (1950).
208. Wolffe, A. H., N. D. Henderson, and G. L. McCallum. "*Salmonella* from Dogs and the Possible Relationship to Salmonellosis in Man," *Amer. J. Pub. Health* **38**, 403 (1948).
209. Geldreich, E. E. and R. H. Bordner. "Fecal Contamination of Fruits and Vegetables During Cultivation and Processing for Market–A Review," *J. Milk Food Technol.* **34**, 184 (1971).
210. Gaub, W. H. "Environmental Sanitation–A Colorado Major Health Problem: A Review of the Problem," *Rocky Mountain Med. J.* **43**, 99 (1946).
211. Krishnaswami, S. K. "Health Aspects of Land Disposal of Municipal Wastewater Effluent," *Can. J. Pub. Health* **62**, 36 (1971).
212. Jepsen, A. and H. Roth. "Epizootiology of *Cysticerus bovis:* Resistance of the Eggs of *Taenia saginata,*" in *Report of the 14th Internat. Vet. Congress* **2**, 43 (1952).
213. Dunlop, S. G., R. M. Twedt, and W. L. Wang. "Salmonella in Irrigation Water," *Sew. Ind. Wastes* **23**, 1118 (1951).
214. Oldham, W. K. "Can Municipal Sewage be Spray Irrigated?" *Water Poll. Control (Canada)* **112**, 29 (1974).
215. Rudolfs, W., L. L. Falk and R. A. Ragotzkie. "Contamination of Vegetables Grown in Polluted Soil. IV. Bacterial Decontamination," *Sew. Ind. Wastes* **23**, 739 (1951).
216. Shooter, R. A., E. M. Cooke, H. Gaya, P. Kumar, N. Patel, M. T. Parker, B. T. Thom, and D. R. France. "Food and Medicaments as Possible Sources of Hospital Strains of *Pseudomonas aeruginosa,*" *Lancet* **1**, 1227 (1969).
217. Kominos, S. D., C. E. Copland, B. Grosiak, and B. Postic. "Introduction of *Pseudomonas aeruginosa* into a Hospital via Vegetables," *Appl. Microbiol.* **24**, 567 (1972).
218. Green, S. K., M. N. Schroth, J. J. Cho, S. D. Kominos, and V. B. Vitanza-Jack. "Agricultural Plants and Soil as a Reservoir for *Pseudomonas aeruginosa,*" *Appl. Microbiol.* **28**, 987 (1974).
219. Christovao, D. d. A., J. A. N. Candeias, and S. T. Iaria. "Sanitary Conditions of the Irrigation Water from Vegetable Gardens of the City of Sao Paulo. II. Isolation of Enteroviruses," *Rev. Saude Publica.* **1**, 12 (1967); *Water Poll. Abstr.* **42**, 681 (1969).

220. Antykov, M. S. "Impact of Viruses in Wastewater on Agricultural Irrigation Fields," *Gig. Sanit.* (USSR) **38**, 110 (1973).
221. Mazur, B. and W. Paciorkiewicz. "Dissemination of Enteroviruses in the Human Environment. I. The Presence of Polio Virus in Various Parts of Vegetable Plants Grown in Infected Soil," *Microbiol.* (Pol.) **25**, 93 (1973).

6

HYDROGEOLOGICAL ASPECTS OF LAND APPLICATION OF WASTEWATER

Don L. Warner

Professor of Geological Engineering
University of Missouri
Rolla, Missouri 65401

INTRODUCTION

Consideration of site hydrogeology is important because, although land disposal systems are intended to improve the quality of the applied effluent, the wastewater that reaches the groundwater system will usually be of poorer quality than the native groundwater. Therefore, the contaminating effect of the applied wastewater must be minimized. Also, when the groundwater table approaches the land surface too closely, it interferes mechanically with the disposal operation. The goals of efficient system operation and minimal groundwater contamination are achieved by proper site selection, system design, and system operation.

HYDROGEOLOGIC EVALUATION OF SITE

Preliminary Evaluation

Groundwater Occurrence

A classification of subsurface water is shown in Figure 6.1, in which groundwater is shown as being in contact with the atmosphere. Groundwater in this state is classed as unconfined. It is subject to direct contamination by infiltrating water, and water levels respond rapidly to local

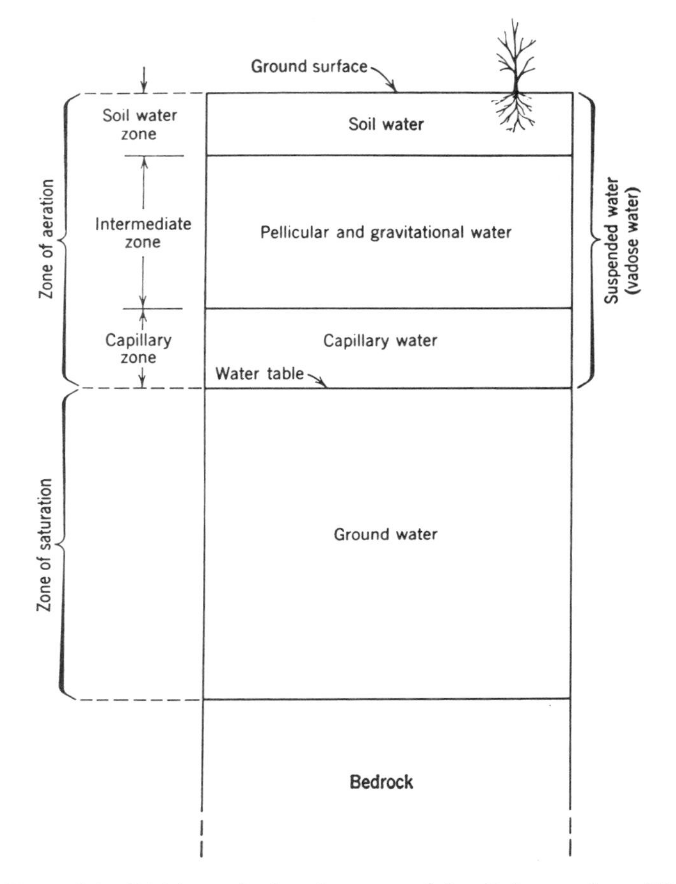

Figure 6.1. Divisions of subsurface water (after Reference 1, p. 18).

recharge. Confined groundwater, which is isolated from the surface by a relatively impermeable bed (aquiclude) consisting of clay, shale, or dense limestone, is not so easily contaminated, nor is it affected much by local sources of recharge. Aquifers that are confined by beds that are not highly permeable but still can transmit a significant amount of water (aquitards) are termed semiconfined.

In defining local groundwater occurrence, the investigator will first want to acquaint himself with the general geology and topography. He should ask, for example, is the site underlain by igneous intrusive or metamorphic rocks, volcanic rocks, or sedimentary rocks; what is the bedrock structure; is there a cover of unconsolidated sediments or soil and if so, how thick and of what type? Preliminary bedrock geologic information can usually be obtained from available maps and reports or by consultation with state agencies, such as state geological or water surveys. The general nature and thickness of unconsolidated sediments and residual soils in an area may also be described in geologic reports or be known to geologists in the area; sometimes, however, it may be necessary to obtain this information from agricultural soil survey reports or from representatives of the federal or state agricultural agencies.

In many cases, available topographic maps will provide the topographic information necessary for a preliminary site evaluation. Aerial photographs are a valuable supplement to topographic maps or can provide topographic data when no suitable maps are available. Surface reconnaissance of a site may be necessary when no maps or photos can be obtained and, if practical, is probably a good practice in any case.

Once the geologic and topographic framework has been established, the investigator will next want to know which geologic units have been found to be aquifers, which are aquicludes or aquitards, and the approximate hydrologic properties of these units. As with geology, the groundwater occurrence in most areas of the United States will be sufficiently well-described in reports by the U.S. Geological Survey or by state agencies to allow a preliminary assessment of groundwater conditions at a proposed site. If the literature is not adequate, discussion with federal and state agency personnel will usually provide the necessary information.

Some unsuitable sites can be eliminated or at least classed as very undesirable as a result of this preliminary evaluation. An example of such a site would be one having relatively impermeable bedrock such as shale, dense limestone, or crystalline igneous rock at or within a very few feet of the surface and having topography such that applied effluent would travel only a short distance through the soil before reappearing and flowing on the surface. Another type of site that could be classed as very poor would be an area of karst topography, where clayey residual soils overlie limestones or dolomites with fracture and solution porosity and permeability. In such locations, infiltration into the soil itself is very slow, but effluent will rapidly enter the bedrock where soil is absent or will wash the thin soil cover into fractures or solution channels, thus developing paths for direct or nearly direct flow of the applied wastewater into the groundwater system. Figure 6.2 depicts this phenomenon. Parizek[2] reported that two large piping holes developed at a land disposal site in such a location with 40 to 60 feet of soil, and that similar but more numerous sinkholes developed at a second site where the soil cover was less than 10 feet.

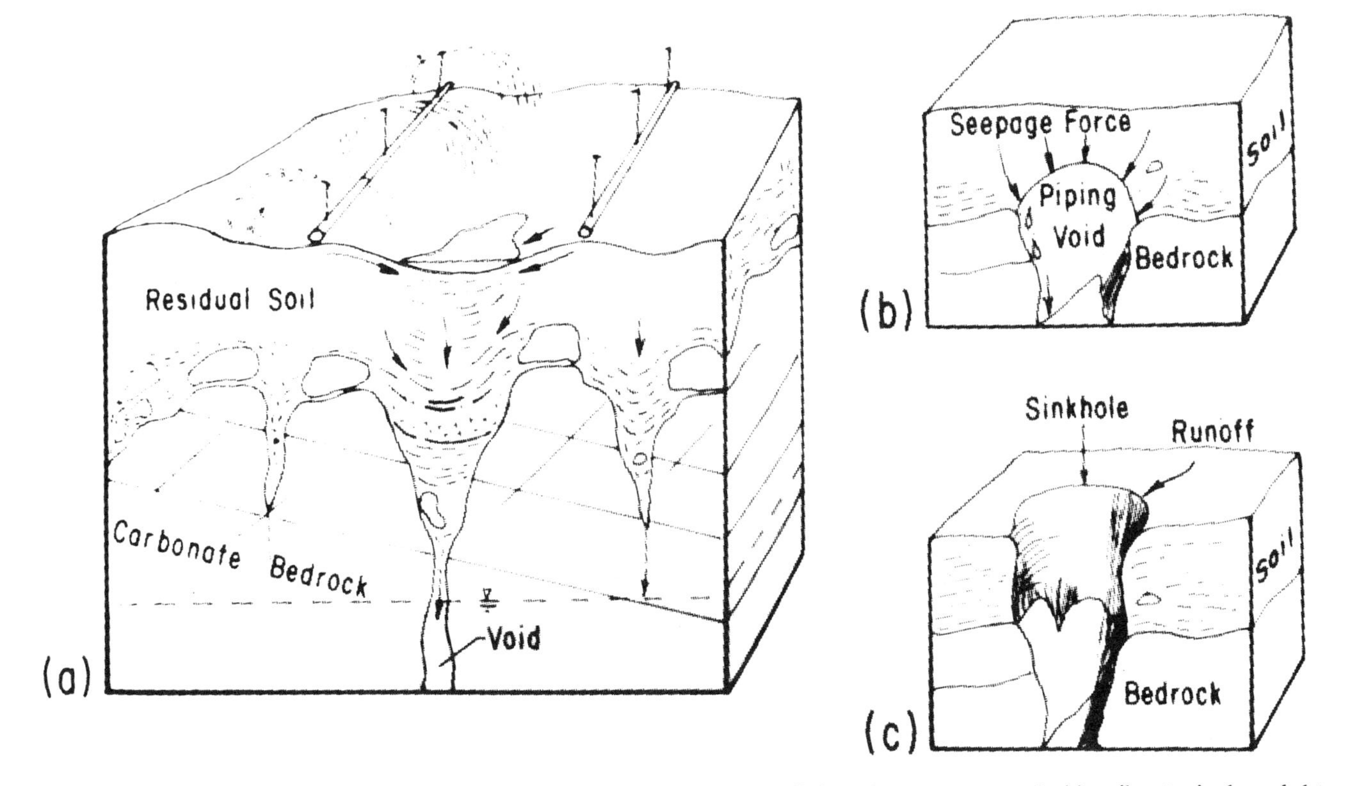

Figure 6.2. Development of piping voids and sinkholes when irrigating residual soils in carbonate terranes. In (a), soil water is channeled to voids in bedrock. Seepage forces help to erode overlying soil into voids (b), and sinkholes may finally develop allowing for recharge of unrenovated effluent (c). (After Reference 2, p. 114.)

In discussing the application of wastewater to the land in the classic karst Ozark region of Missouri, Howe[3] stated:

> Nonconsumptive irrigation of sewage plant effluents in the Ozark region is not in consonance with (1) prevention of deterioration of water quality both at the surface and underground, or (2) consideration of the long-term economic consequences of such deterioration. Groundwater in the Ozarks is particularly vulnerable to contamination. Unless irrigation is carefully practiced and restricted to consumptive applications, irrigated treatment plant effluent will inevitably percolate downward to the saturated zone. From there it will migrate until it: (1) seeps into a stream that transects the water table, (2) is discharged to surface drainage at a spring, (3) is pumped from a well (regardless of its construction) or (4) enters, along with other regional recharge water, the as yet unmeasured groundwater reservoirs that are so important to the Ozark region.

A third type of site that could be immediately classed as undesirable would be one with very little topographic relief, where the groundwater table is at or very near the surface. In extreme situations, such sites may be subject to periodic flooding or may be groundwater discharge areas. The problem is, of course, that applied effluent cannot receive the benefits of treatment by the soil if it is prevented from infiltrating because the soil is already saturated. It might be feasible to provide artificial drainage in the form of the tile drains, ditches, or wells to keep the water table far enough beneath the surface to permit irrigation. However, such management would be expected to add considerably to the cost of a land disposal project.

Groundwater Quality

Since groundwater quality can be expected to be affected by land disposal in most cases, a knowledge of the natural water quality in the area is mandatory. As will be discussed later, the effect of application of wastewater on the quality of the existing groundwater can be estimated if the qualities of both waters and the site characteristics are known. During preliminary site investigation, the general quality characteristics of the groundwater can usually be obtained from the sources previously mentioned.

Groundwater Use

Interpretation of the effect of a disposal project on quality characteristics should be tied to existing and anticipated future uses. For example, an increase in total dissolved solids does not equally affect all water uses. Water with total dissolved solids contents below 500 mg/liter is of good quality for drinking and for irrigation of all types of crops. As the level of dissolved solids increases, water becomes increasingly less desirable for both uses; beets or cotton, however, would be little affected by TDS contents of 1000 mg/l, while this level of salinity would be damaging to citrus fruits. Addition of nitrate to groundwater used for domestic purposes is undesirable, but added nitrate may be beneficial where groundwater is used solely for irrigation. Detailed criteria for the quality of water for various uses, such as those developed by the Federal Water Pollution Control Administration,[4] should be consulted in order to evaluate the influence of a particular quality change on a specific use.

During the preliminary site investigation, present users of groundwater in the vicinity of the site should be identified. The radius of investigation depends on the groundwater occurrence as previously determined and the size of the disposal project. It would also be prudent to inquire into patterns of land ownership and probable future use of the groundwater to avoid, if possible, eventual political or legal controversy over the effects of the disposal operation on groundwater resources.

Detailed Site Evaluation

If conditions are sufficiently promising after the preliminary site evaluation, a more detailed site investigation will normally follow. The exact procedure for such an investigation cannot be outlined because each site will be sufficiently different to require individual consideration. Generally, the objective will be to establish more accurately the same characteristics examined in the preliminary evaluation. However, perhaps some of these will be sufficiently known and no further effort will be needed.

Groundwater Occurrence

To determine the local geology and groundwater occurrence and quality in detail, test holes are nearly always needed. These can range in type from backhoe trenches or hand auger holes to large diameter test wells constructed with rotary drilling equipment, depending on the site and the objective of exploration. Shallow exploration would be for study of the upper few feet of soils, which are most important in the treatment of the applied effluent. A drilled well or wells might be necessary for determination of aquifer properties, for water sampling, and for monitoring of groundwater conditions during operation.

The properties of soil and unconsolidated clastic sediments important in a detailed hydrogeological investigation include grain-size distribution, mineralogy, clay content, ion exchange capacity, moisture content, porosity, and permeability. These properties determine the rates of infiltration and lateral movement, water storage capacity, filtering capacity, and capacity for retention of chemicals by sorption or reaction. At a site where the mantle of unconsolidated material is relatively thick (probably greater than 50 feet) and where the water table lies within this material, bedrock geology may be of only minor importance. At other sites, where the water table lies in a bedrock aquifer, its properties require detailed investigation. Sandstone aquifers are lithified sand aquifers, and thus have similar properties though perhaps in different amounts. Limestones, dolomites, volcanics, and igneous intrusive rocks are all generally much different than clastic sediments or sandstones in that their porosity and permeability is usually in the form of fractures or, in carbonates, solution channels. This type of porosity and permeability does not provide the filtration or sorption found in sands, silts, or sandstones, and once a polluted water enters an aquifer with fracture or solution porosity, it often travels rapidly with little if any further treatment.

An important exception to this generalization is that limestone or dolomite aquifers react with an acidic waste resulting in an increase in pH. A secondary benefit of this neutralization is precipitation of ions, particularly metals, that are less soluble as pH increases.

A variety of field and laboratory methods are available to assess the hydrogeologic properties of soils, sediments and rocks. Field methods include surface and borehole geophysical measurements,[5,6] infiltration tests for permeability measurement in unsaturated soils,[7] and pumping tests for evaluation of aquifer properties.[8] Comprehensive discussion of laboratory chemical analysis of soils is provided in a manual by Jackson[9] and laboratory physical soil tests are outlined by Lambe.[10]

As a result of test holes, and perhaps surface geophysical surveys, maps of the site can be constructed showing the thickness and character of surficial materials and of the bedrock geology, if applicable. The water table surface can be represented by a contour map referenced to an absolute datum and also by a contour map of depth from the ground surface. Both groundwater maps have important uses. The former is a groundwater equipotential map, and lines drawn perpendicular to the equipotential lines are approximate flow lines, as exemplified by Figure 6.3, in which areas A and E are recharge areas. At A, recharge is from the stream; at E, recharge is by infiltration of surplus irrigation water, and a groundwater mound has formed. Area D is a groundwater depression, perhaps caused by pumping. At C, groundwater is being discharged to the stream. Large changes in aquifer permeability cause convergence (increase of permeability) or divergence (decrease of permeability) of flow lines.

Figure 6.4 is a depth-to-water map of the same areas shown in Figure 6.3. This map may have particular application in a land disposal operation

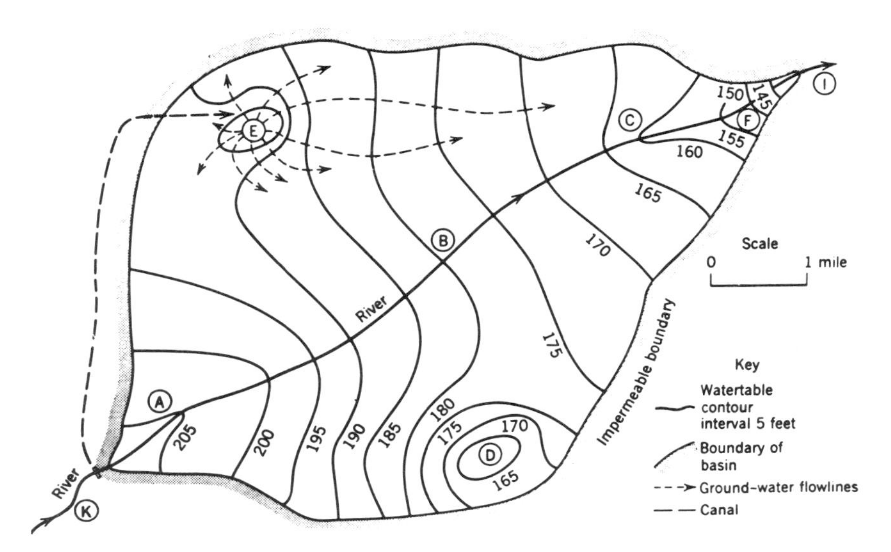

Figure 6.3. Contour map of the water table in a small hypothetical groundwater basin. If the aquifer is homogeneous and isotropic and if the slope of the water table is not large, the map can be used to construct a flow net, *i.e.*, a regular "square" net. A small number of flowlines have been drawn on the map. Excessive convergence of the flowlines suggests a changing transmissivity of the aquifer. (After Reference 11, p. 49).

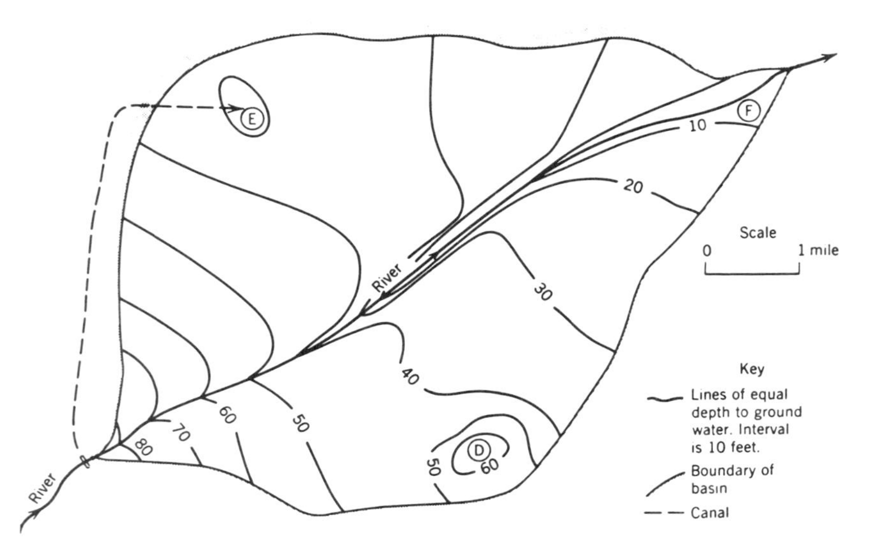

Figure 6.4. Depth-to-water map of the same groundwater basin shown in Figure 6.3. (After Reference 11, p. 51).

where control of the depth to water is important in maintaining maximum possible treatment of applied effluent.

It should be noted that water levels in shallow aquifers are often subject to great seasonal variation, and water level observations should be made over a sufficient length of time to define the natural pattern of rise and fall.

Water Quality

During a detailed site investigation, samples should be obtained from test wells, water supply wells, springs, streams, and lakes in the immediate area. All waters potentially affected should be sufficiently well-characterized so that any changes caused by the future land disposal operation can be recognized. Groundwater tends to maintain a rather constant quality in the absence of man-caused changes, so that repeated sampling is usually not necessary to characterize it. Surface waters may need to be sampled periodically over time to document natural seasonal quality changes and variations caused by intermittent sources of influence. The quality parameters selected for measurement should, of course, be based on the constituents in the applied wastewater.

Water Use

Where significant present or projected future uses of groundwater have been found in the vicinity of the proposed site, the locations of the discharge points and the amounts removed should be determined. The quality requirements of the users should also be established so that any undesirable effects of land disposal on water users can be anticipated.

Other Factors

In order to incorporate hydrogeologic factors into the design of a land disposal system, other site characteristics may need to be known: the annual amount and seasonal distribution of precipitation and evaporation, temperatures at the site, and the type and distribution of vegetation and its transpiration characteristics.

HYDROGEOLOGY IN SYSTEM DESIGN

Hydrogeology enters into the system design in two separate but related ways. First, the effect of application of wastewater on groundwater levels and flow patterns at the site and in the vicinity need to be determined. This leads to the selection of application rates and patterns, which is accomplished by analytical equations or models, generally numerical or electrical, based on the Darcy law of flow through porous media. For example, Bittenger and Trelease[12] developed an analytical equation to describe the development and dissipation of a groundwater mound beneath a recharge basin. Bouwer[13] used an electric analog model for analysis of recharge mounds, and Bouwer[14] provides examples of the use of both electrical models and analytical equations in the design and operation of the Flushing Meadows Project, a pilot project for groundwater recharge by high-rate application of municipal sewage effluent.

In recent years, rather sophisticated computer models have been developed for use in groundwater studies.[15,16] These models are capable of incorporating the many variables that may influence groundwater levels in a large land disposal project, including the lateral and vertical variations in hydraulic properties that are usually found.

Analysis of changes in groundwater quality resulting from wastewater application is the other major hydrogeological design problem. Quality changes are, of course, related to the volume of wastewater reaching the groundwater system, so this design step is a continuation of the analysis of water level and flow pattern changes described above. Additional factors that may need to be incorporated into an analysis of quality changes are: (1) dilution of the waste by mixing with unpolluted water, and (2) removal of pollutants by sorption, biological action, chemical reaction, radioactive decay, or uptake by vegetation.

Clearly, accounting for such factors can become very complex. In some situations, it may be satisfactory for an experienced hydrogeologist to make an estimate based on a greatly simplified conceptual model of the hydrogeologic system. An estimate of this type can be very useful in establishing the extreme range of possible results. However, if the results of assuming different values for the many possible variables must be tested, a relatively sophisticated model is required, probably one using a digital computer. Although no example is known to the author where computer modeling has been applied to hydrogeology in a land disposal operation, the ability to perform this type of analysis is available when the problem justifies the effort.

MONITORING

Hydrogeologic monitoring of land disposal operations should be used to define changes in groundwater levels or changes in the quality of groundwater and surface waters affected by groundwater. Observation wells are necessary for accurate determination of groundwater level changes. No specific recommendations can be made for the number and location of observation wells, which must be selected on the basis of the hydrogeology and size of the particular site and on the degree of concern about water levels. If the groundwater table is very deep and there is no concern about its level from an operational view, then monitoring of a single well at the point of greatest anticipated change might be sufficient. It must be remembered, however, that flow rates and flow directions are dependent on water levels, and the data needs for this purpose may control the number and location of observation points. Also, if wells are constructed for water quality sampling, the water levels in these wells should be measured as an aid in interpreting any quality changes observed.

The monitoring of water quality at a land disposal site should, as previously discussed, begin during site evaluation when the background quality of ground and surface waters is established. Next, predictions of quality changes should be made in the design phase. Monitoring during operation is needed to meet regulatory requirements, to check design predictions, and to provide data for modification of system design and operating procedures, if necessary.

Some means of surveillance of groundwater quality during the operation of a land application operation are: water sampling, measurement of groundwater levels, geophysical measurements, remote sensing, and maintenance of material balances. Possible sampling points are:

1. *Monitor wells.* These may be specially constructed or may be existing wells converted for use as monitor wells. A monitor well may be pumped or unpumped. An unpumped well samples only the water that passes directly through the well bore. A pumped monitor well produces an integrated sample from an area whose size depends on the local geohydrology and the rate of pumping.
2. *Water supplies.* Samples of groundwater pumped for water supply can be periodically taken and analyzed to detect quality changes. Such samples are representative of the well or wells in the system and the area or areas of influence.
3. *Springs.* Since springs are outlet areas for groundwater, samples taken from them are similar to those from pumped wells in that they reflect the quality of water within an area of influence.
4. *Streams.* Because many streams derive most of their flow from groundwater drainage for a substantial part of each year, samples from gaining streams can be used to measure groundwater quality. Similarly, samples from losing streams can yield information on pollution entering the groundwater from surface water sources. A gaining stream might be thought of as representing the composite quality of a number of springs.

Location of sampling points is most important in obtaining the desired results from a monitoring program. As LeGrand[17] has stated, haphazard plans for the location of monitoring sites are almost certain to result in excessive cost and to fail in their objective. He further points out that such sites must be located on the basis of the hydrogeologic framework, and he provides some useful general guidelines for planning their location.

The frequency of sampling at a selected monitoring point depends upon the sensitivity of the quality at the site to natural and man-caused influences. Under natural conditions the quality of groundwater will typically change imperceptibly with time. Rates of change are related to rates of flow, which in turn are governed by the hydrogeologic situation. Some groundwater basins unaffected by man show annual fluctuations in quality produced by seasonal variations in recharge, level changes, and discharge. Common patterns of man-caused change are an increase in amplitude over periodic natural fluctuations, rapid local change in response to a new pollution source, and gradual progressive deterioration in quality over an extended period of time. Where rapid transfer of pollutants to the groundwater and rapid movement in the groundwater system are a possibility, daily or weekly sampling may be appropriate.

To characterize gradual changes, monthly or quarterly sampling should be adequate. Sampling during the early life of a land disposal operation should generally be at maximum frequency. Once a pattern of movement of pollutants has been established a schedule of less frequent sampling can be adopted. At some sites, movement of pollutants through the unsaturated zone may be very slow and, at such locations, sampling of that zone may be required.

The analyses performed on the collected samples should be carefully selected to focus on specific potential pollutants known to be in the applied effluent because of their hazard, persistence, concentration, ease of identification, or other characteristic features.

Although measurement of groundwater levels does not provide quality data directly, it can yield valuable indirect information. As previously shown, the groundwater flow pattern of a region can be defined from a map of water level contours. Changes with time of flow patterns due to wastewater application may be apparent and may allow deduction of the direction of flow of pollutants.

Geophysical measurements made in bore holes or on the ground surface can provide valuable supplementary information in a monitoring program. A wide variety of bore hole logging methods have been developed. Keys and MacCary[6] provide an extensive discussion of the use of bore-hole geophysics in water-resource investigations.

Bore-hole geophysical devices may be designed to examine only the fluids in the bore hole or to examine a volume of the aquifer around the bore hold. Two fluid properties directly measured in the bore hole are temperature and conductivity. Two aquifer properties commonly measured are conductivity (resistivity) and natural radioactivity, which contribute to the overall reading, and changes in which can be detected even though the instruments view a section of the aquifer rather than the water alone.

Electrical resistivity has been widely used to delineate areas of polluted groundwater. A recent article by Stollar and Roux[18] cites several case histories and lists the publications of previous investigators. Resistivity surveys depend on detection of the existence of a contrast between the resistivity of uncontaminated and contaminated groundwater. Therefore, for the method to be useful at a particular site, the contrast must be greater than the variation in values caused by geologic factors.

Remote sensing is the technology of remotely collecting and interpreting data generated by electromagnetic energy from the earth's surface and near surface. This definition includes conventional black and white photography as well as the many newer methods that utilize other parts of the electromagnetic spectrum. Barr and James[19] summarized many of the uses of remote sensing including aerial monitoring of surface water quality. In most cases monitoring of groundwater quality will probably be indirect. Remote sensing has been widely used for agricultural studies and has been found useful for detection of drainage and salinity problems.[20] This suggests the use of remote sensing for indirectly inferring groundwater quality problems based on soil conditions and on the response of vegetation to changes in groundwater quality.

In land disposal operations, the soil and vegetation are expected to remove chemicals that would otherwise enter the groundwater. A means of measuring the effectiveness of the soil and vegetation in performing this task is the material balance. If much more of a chemical is being applied than is being retained, then the remainder may be entering the groundwater system. The material balance for nitrate is complicated by the fact that nitrate may be reduced by bacterial action to nitrogen gas, which escapes into the atmosphere. Surface runoff of part of the applied effluent also complicates such an analysis.

REFERENCES

1. Todd, D. K. *Ground Water Hydrology.* (New York: John Wiley & Sons, Inc., 1959).
2. Parizek, R. R. "Site Selection Criteria for Wastewater Disposal–Soils and Hydrogeologic Considerations," in *Recycling Treated Municipal Wastewater and Sludge Through Forest and Cropland,* W. E. Sopper and L. T. Kardos, Eds. (University Park, Pa.: Pennsylvania State University Press, 1973), pp. 95-147.

3. Howe, W. B. "Effluent Disposal by Irrigation in the Ozarks," *Missouri Mineral News* **12**(12), 214 (1972).
4. Federal Water Quality Control Administration. *Water Quality Criteria.* (Washington, D.C.: U.S. Department of the Interior, FWPCA, 1968).
5. Zohdy, A. A. R., G. P. Eaton, and D. R. Mabey. "Application of Surface Geophysics to Ground-Water Investigations," *Techniques of Water-Resources Investigations of the United States Geological Survey,* Book 2, Chapter D1, U.S. Geological Survey (1974).
6. Keys, W. S. and L. M. MacCary. "Application of Borehold Geophysics to Water-Resources Investigations," *Techniques of Water-Resources Investigations of the United States Geological Survey,* Book 2, Chapter E1, U.S. Geological Survey (1971).
7. Bouwer, H. "Field Determination of Hydraulic Conductivity Above a Water Table with the Double-Tube Method," *Proc. Soil Sci. Soc. Am.* **26**, 330 (1962).
8. Ferris, J. G., *et al.* "Theory of Aquifer Tests," U.S. Geol. Survey Water-Supply Paper 1536-E (1962).
9. Jackson, M. L. *Soil Chemical Analysis* (Englewood Cliffs, N.J.: Prentice-Hall, Inc., 1958).
10. Lambe, T. W. *Soil Testing for Engineers.* (New York: John Wiley & Sons, Inc., 1951).
11. Davis, S. N. and R. J. M. DeWiest. *Hydrogeology.* (New York: John Wiley and Sons, Inc., 1966).
12. Bittenger, N. W. and F. J. Trelease. "The Development and Dissipation of a Ground-Water Mound Beneath a Spreading Basin," ASAE Paper No. 60-708, presented at the 1960 Winter Meeting Am. Soc. of Agricultural Engrs., 1960.
13. Bouwer, H. "Analyzing Ground Water Mounds by Resistance Network," *Proc. Irrigation and Drainage Div. Am. Soc. Civil Eng.* **88**, IR3, 15 (1962).
14. Bouwer, H. "Renovating Secondary Sewage by Ground Water Recharge with Infiltration Basins," U.S. Environmental Protection Agency Water Pollution Control Research Series Report 16060 DRV 03/72 (1972).
15. Prickett, T. A. and C. G. Lonnquist. "Selected Digital Computer Techniques for Groundwater Resource Evaluation," Illinois State Water Survey Bulletin 55, Urbana, Illinois (1971).
16. Pinder, G. F. "A Digital Model for Aquifer Evaluation," *Techniques of Water Resources Investigations of the United States Geological Survey,* Book 7, Chapter C1, U.S. Geological Survey (1970).
17. LeGrand, H. E. "Monitoring of Changes in Quality of Ground Water," *Ground Water* **6**(3), 14 (1968).
18. Stollar, R. L. and P. Roux. "Earth Resistivity Surveys–A Method for Defining Ground-Water Contamination," *Ground Water* **13**(2), 145 (1975).
19. Barr, D. J. and W. P. James. "Application of Remote Sensing in Civil Engineering," Am. Soc. of Civil Engrs. Environmental Engineering Meeting, Oct. 29-Nov. 1, 1973, New York, N.Y., Meeting Preprint 2072 (1973).
20. Meyers, V. I., L. R. Ussery, and W. J. Rippert. "Photogrammetry for Detailed Drainage and Salinity Problems," *Trans. Amer. Soc. Agric. Eng.* **6**(4), 332 (1963).

7

LAND DISPOSAL OF LIQUID INDUSTRIAL WASTES

A. T. Wallace

Professor, Department of Civil Engineering
University of Idaho
Moscow, Idaho 83843

INTRODUCTION

The literature contains a great deal of information on land application of liquid wastes. However, a broad background in science and engineering is required of a person who intends to make any sense out of this information. Questions relative to removal mechanisms, compatibility of waste and soil, hydrogeology, costs, effects of vegetation and cover crops, climatology, health aspects and many other areas are vital to the success of a system and must be carefully considered in connection with each intended application of the technique. Most of these questions are addressed by the other chapters of this book. This chapter concentrates primarily on application of liquid industrial wastes to land. In accomplishing this objective, the following points have been chosen for consideration:

1. the reasons for the attractiveness of the land treatment and disposal alternative to industrial processors,
2. the variety of industrial effluents that have been successfully disposed of on the land,
3. the hydraulic and organic loadings that have been used in existing systems,
4. the possibilities for a quantitative approach to determination of the land area required for a specific application, and
5. the areas where research is needed to aid in improving existing design techniques.

REASONS FOR CONSIDERATION OF THE LAND TREATMENT AND DISPOSAL ALTERNATIVE

When evaluating the alternatives available for treatment and disposal of industrial effluents, many factors must be considered. A summary of those factors that may lead an industry to choose land disposal follows.

1. The proximity of manufacturing operations to rural areas that provide close access to the required acreage. A second consideration related to the rural location is the inability of the smaller local governments to provide the degree of treatment required of industry even if they wanted to do so. In several recorded cases, wet process industries located in small towns have pumped their effluents several miles to reach suitable land disposal sites rather than pay the high cost of conventional treatment with domestic wastes.

2. The seasonal aspect of many of the wet process industries that use the technique. In particular, this includes canning and other food processors and to a certain extent milk and cheese producers.

Of great significance is the seasonal nature of the effluent disposal problem *per se.* The best season for land disposal is often the poorest time for waste disposal in receiving waters because of low flows, high temperatures and seasonal recreational uses of the waters. Coincidently, the low stream flow and maximum demand for irrigation water occur together. This naturally leads to favorable consideration of effluent reuse for irrigation.

3. The rising cost of conventional waste treatment facilities together with increasingly restrictive governmental water quality and effluent discharge regulations have given major impetus to consideration of land disposal. In an example from the author's experience, no portion of the flow of a nearby stream could be allocated for waste assimilation. The cost of a spray irrigation system was estimated as $153,000 less than the cost of an activated sludge system with final disposal through a spreading basin for a 1 mgd flow. The actual cost saving was even larger than estimated.[1]

4. Another advantage of land disposal, related to (3) above, is the fact that land disposal systems do not generally come under the jurisdiction of the U.S. Environmental Protection Agency NPDES permit authority. Thus, the state regulatory agency with jurisdiction will normally be the highest level of bureaucracy involved in the project. This generally insures a more reasonable approach to plan review, standards setting and monitoring requirements; most industries have become painfully aware of this fact in recent years.

5. Increasing intensiveness of agricultural practices in areas that formerly had ample rainfall during the growing season has created a demand

for supplemental irrigation water to take full advantage of the agricultural advances in crop production.

6. Finally, the increasing emphasis on groundwater recharge as a water resources management tool, together with increased knowledge in the areas of wastewater renovation by the soil medium and groundwater flow systems, has prompted more consideration of the technique.

REVIEW OF TYPES OF INDUSTRIAL EFFLUENTS WHICH HAVE BEEN TREATED ON LAND

The general criteria for judging suitability of a liquid effluent for a land disposal application are as follows:

1. The organic material must be biologically degradable at reasonable rates.
2. It must not contain materials in concentrations toxic to soil microorganisms. Since some toxic materials may accumulate through adsorption or ion exchange and approach toxic levels after prolonged operation, there must be reasonable assurance that this effect can either be prevented or mitigated.
3. It must not contain substances that will adversely affect the quality of the underlying groundwater. In many instances, decisions relative to this aspect of land disposal systems are difficult because of the uncertain nature of available estimating techniques.
4. It must not contain substances that cause deleterious changes to the soil structure, especially its infiltration, percolation, and aeration characteristics. An imbalance of sodium ion is a common problem in this regard.

Table 7.1 contains a fairly complete listing of the kinds of industrial effluents that have been disposed of on the land. The references given are those which, in the opinion of the writer, provide the best case histories of specific projects. The interested reader may find these references useful in planning projects for similar wastes. Domestic and animal wastes have been purposely omitted from this table as they are the subject of other chapters.

As can be seen from the table, a diverse spectrum of wastes have at one time or another been subjected to land disposal. Although most of the references tabulated give details on somewhat successful systems, many failures are also recorded, and these provide useful insight into use of the land disposal technique.

Table 7.1. Summary of Types of Industrial Effluents Disposed of on Land

Industry	References[a]
Food processing	
Canning, frozen and convenience foods	
Vegetables, including baby food	2,3,4,5,6,7
Soup	8,9
Fruit, except citrus	7,10
Citrus fruit	11,12
Instant coffee and tea	13
Dairy products	
Milk plants	3,14,15
Cheese (whey)	16,17
Meat products	18,19
Miscellaneous	
Potato processing	1
Sugar beets	
Starch plants	
Wine	
Distilling (spirits)	
Pulp and Paper	20,21
Sulfite	22,23,24,25
Kraft (sulfate)	25,26,27
Semi-chemical	28
Strawboard	29
Groundwood	
Hardboard and insulation	30,31
Boxboard and paperboard	32
Deinking	33
Miscellaneous	
Tanning	34
Textiles	
Pharmaceuticals	35
Biological chemicals	36
Explosives	37
Wood distillation	38
Rope and hemp	
Coal-tar chemicals	
Cooling water from aluminum casting	39

[a]Where no reference is given, only limited performance data are available.

WASTE LOADINGS

Although many investigators have stated that design of land application systems will always be controlled by the hydraulic loading, the evidence is overwhelming that either hydraulic or organic loading can control in a given instance. In some rare cases, neither will be as important as the cation loading, especially with regard to sodium. In a svstem the author has recently designed, nitrogen loading dictated the land requirement. In general, however, for wastes with BOD concentrations in the 1000 mg/l or less range, hydraulic loading will usually control the design.

The descriptions of operating systems in the literature seldom report enough analytical data from which meaningful correlations can be made. For example, descriptions of soil characteristics, if reported at all, are usually limited to type and texture. More information is needed on the physical and chemical characteristics of the soils that may play a role in the removal phenomena. More will be said about this later. Meanwhile, it seems appropriate to summarize the hydraulic and organic loading rate data that are available in the literature (Table 7.2).

Clearly, if sufficient data were also available on the soil and waste characteristics and on the quality of the percolates in all the above cases, it might be possible, through an examination of all the variables, to provide some sort of a generalization of all the data. This might take the form of either a greatly simplified mathematical model of the soil system or, more likely, an empirical expression relating performance to the variables. At the present time, however, not enough systematic studies have been undertaken to provide a basis for either approach.

POSSIBILITIES FOR A QUANTITATIVE APPROACH

There have been a few major studies that do provide some insight into possible techniques for providing a rational approach to design. In all cases, these studies have involved the use of lysimeters to obtain the necessary data under conditions sufficiently well-controlled to be meaningful. Lysimeters can take many forms. One type, which can be constructed from a length of 36-inch (91.4-cm) diameter concrete or clay pipe, is shown in Figure 7.1.

Lysimeter studies conducted by the National Council for Stream Improvement using unbleached kraft and kraft black liquor[20-21] and by the Wisconsin State Committee on Water Pollution using waste sulfite liquor[22] have produced enough data under controlled conditions to give rise to the kind of plots shown in Figure 7.2.

In two of the studies[21-22] the lysimeters were operated with the media unsaturated so the action was primarily aerobic. However, in one of the

Table 7.2. Summary of Hydraulic and Organic Loading Rates Used in Existing Land Disposal Systems for Industrial Wastes

Type of Waste	Hydraulic Load gal/ac day[a]	Organic Load lb BOD/ac day[b]
Biological chemicals	1,500	370
Fermentation beers	1,350	170
Vegetable tanning		
Summer	54,000	360
Winter	8,100	54
Wood distillation	6,850	310
Nylon	1,700	287
Yeast water	15,100	–
Insulation board	14,800	138
Hardboard	6,000	85
Boardmill whitewater	15,100	38
Kraft mill effluent	14,000	26
RI[c]	350,000	120
Semichemical effluent	72,000	90-210
Paperboard	7,600	13-30
Deinking	32,400	108
Poultry	40,000	100
Peas and corn		
57 day pack	49,000	238
35 day pack	34,400	2,020
Dairy		
Low value	2,500	10
High value	30,000	1,000
Soup	6,750	48
Steam peel potato	19,000	80
Instant coffee and tea	5,800	92
Citrus	3,100	51-346
Cooling water – aluminum casting (RI)	95,000	35

[a]Multiply by 9.35×10^{-3} to convert to m^3/ha day.
[b]Multiply by 0.89 to convert to kg/ha day.
[c]RI – Rapid Infiltration.

National Council studies[20] the lysimeters were operated under saturated conditions, thus simulating an anaerobic soil environment. Although the wastes used in these studies were not identical, they were similar in their content of wood sugars, hemi-celluloses and cellulose. Therefore, these data are roughly comparable and serve to illustrate the first important point: aerobic degradation rates are much higher than anaerobic rates.

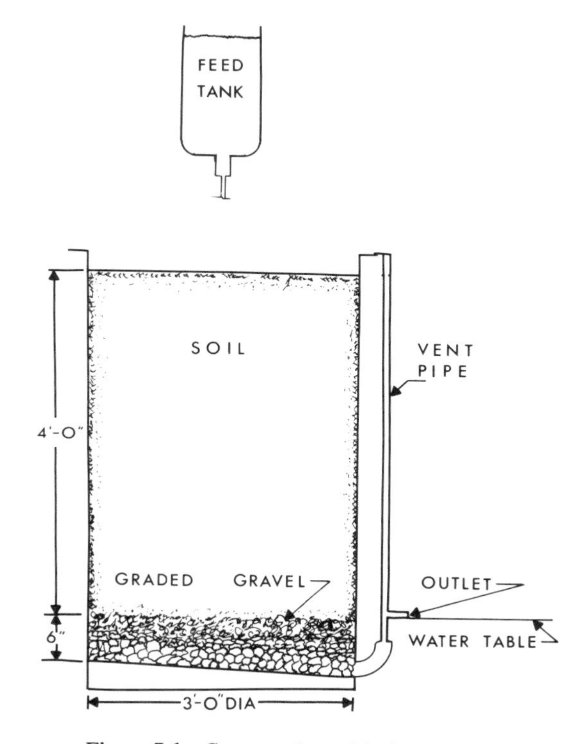

Figure 7.1. Cross section of lysimeter.

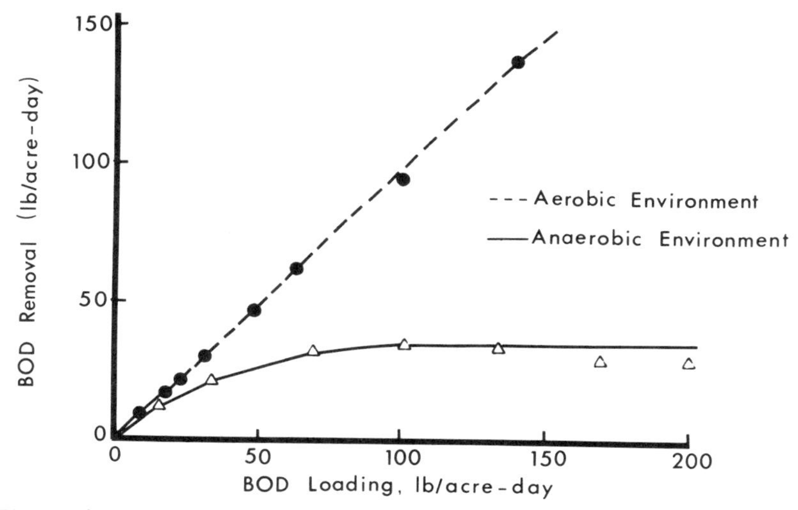

Figure 7.2. Comparative BOD removal rates in aerobic and anaerobic land treatment of wood pulping wastes.

Another interesting aspect of land disposal of industrial wastes is shown in Figures 7.3 and 7.4. Figure 7.3, based upon the work of the forest ecologist Minderman,[40] depicts the relative decomposition rates of various groups of chemical constituents found in forest litter. It shows clearly how the decomposition rates are roughly related to the chemical complexity and solubility of the various components. Figure 7.4 shows a comparison of Minderman's data on cellulose decomposition with data on cellulose degradation obtained in other studies under diverse environmental conditions. It must be pointed out that these curves are not comparable in the strict scientific sense because of the possible differences in the physical nature of the cellulose fractions and the tremendous number of environmental factors that were not controlled in the studies producing these data. Also they are drawn as assumed first-order rate curves; a more careful inspection of the data would reveal that the reactions start out as first-order and then become higher order as the physical state of subdivision

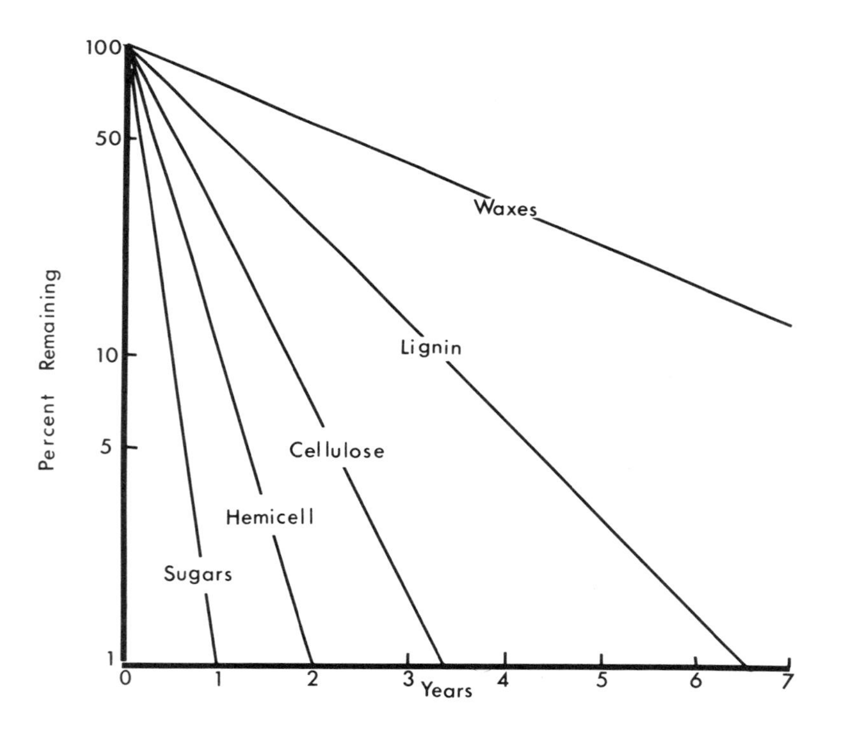

Figure 7.3. Decomposition curves for various organic constituents in forest litter. (After G. Minderman[40])

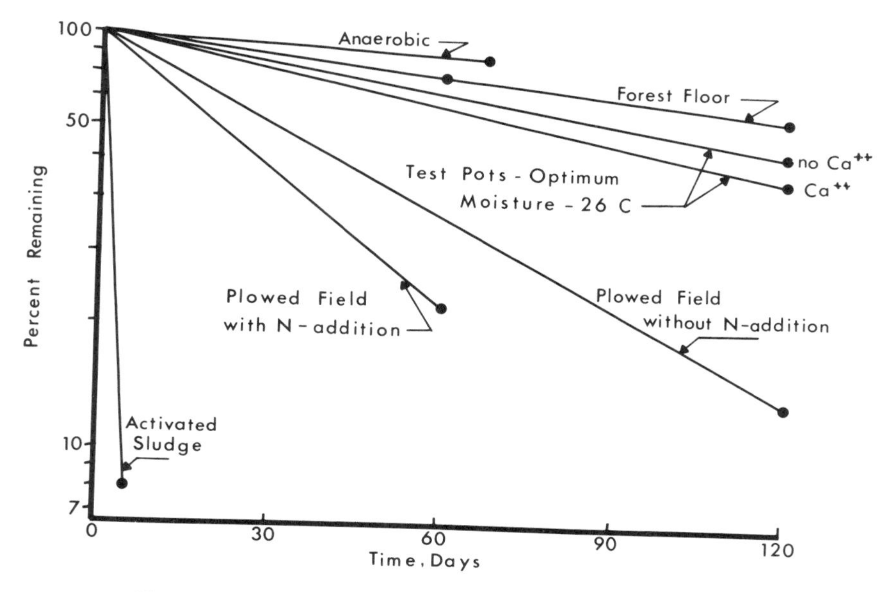

Figure 7.4. Rates of cellulose decomposition in soil and water under various environmental conditions (after several investigators).

becomes greater. The primary usefulness of the curves as drawn is in dramatizing the influence of environment on any decomposition process in the soil.

ON A RATIONAL MODEL

Based upon the data presented thus far, an impression may have been formed regarding the impracticality or perhaps even impossibility of formulating a rational design equation for land disposal processes. This is a somewhat similar situation to that which confronted the National Research Council in their attempts to generalize from the waste treatment data collected at U.S. military installations during World War II.[41] (The reader may recall that efficiency data for biological processes, correlated against rational loading parameters, produced a very poorly defined relationship, somewhat akin to a shot pattern from a shotgun. Nevertheless, the NRC committee settled on a trend line for each, which became the basis for the NRC formulas, the most used of which is the one for trickling filters.) Our problem is even less tractable than theirs because of our inability to classify our systems into groups having similar physical characteristics. That is, the physical, chemical and biological properties of soils are more diverse and therefore less amenable to analysis than are the equivalent properties of trickling filters or activated sludge systems.

There are several points that suggest that the situation is not entirely hopeless. Consider a more detailed plot (Figure 7.5) of the data from the aerobic (unsaturated) lysimeters shown previously in Figure 7.2, extended to include much higher organic loadings. These data seem to imply a high but definitely limited capacity for removal of BOD by a soil column. This upper limit may be related to several factors but certainly the most important of these is the oxygen supply if all other conditions remain roughly equal.

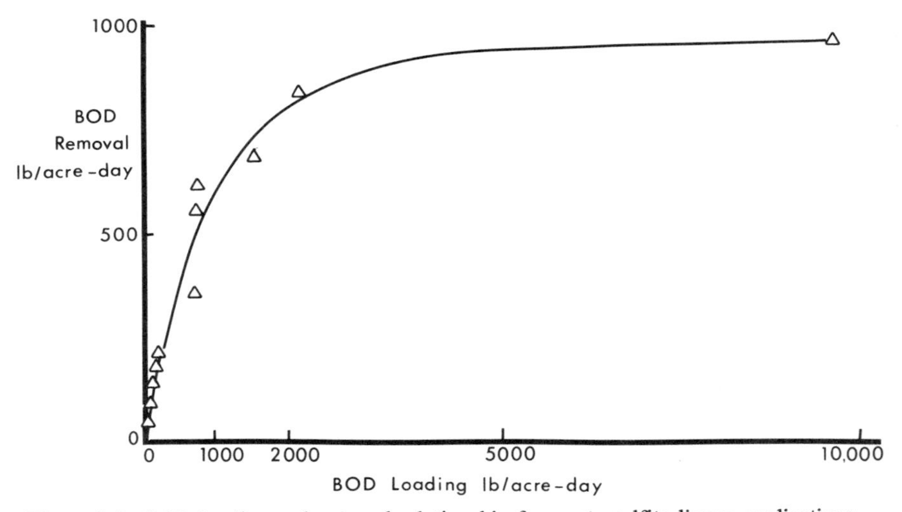

Figure 7.5. BOD loading and removal relationship for waste sulfite liquor applications to 10-foot deep soil columns (after Wisniewski, *et al.*).[22]

For example, in the study from which most of the above data were obtained,[22] it was noted that at the highest BOD loading (9,600 lb/acre-day) the wood sugars were 100% decomposed in the percolates. However, the overall BOD reduction was barely 10%. This simply indicates a shift from aerobic to anaerobic metabolism. Judging from the shape of this curve, this shift probably begins at a loading of 1000 lb/acre-day or so. The only way to define it accurately could be to measure oxygen tension in the soil as a function of loading. Should a question arise as to the drastic difference between the maximum BOD removal shown in Figure 7.5 and the maximum BOD removal shown in the lower curve in Figure 7.2, it should be pointed out that the anaerobic condition in the study of Figure 7.2 was caused by waterlogging the soil and not by overloading it with organic matter, as in Figure 7.5.

Costopoulos[42] investigated the oxidation rate of organics from skim milk solids in closed soil systems. He used hydraulic and organic loadings, time and number of loading cycles as variables, keeping temperature and soil moisture constant. Two mathematical models were generated, one for periodic loading and one for continuous loading. The equations, written for the equilibrium BOD content of the soil solution, are given as:

for periodic loading,

$$L = L_o \exp(-knT) + \ell \frac{1 - \exp[-k(n+1)T]}{1 - \exp(-kT)} \quad (1)$$

for continuous loading,

$$L = \frac{l_o}{k} + (L_o - \frac{l_o}{k}) \exp(-kT) \quad (2)$$

where:

L_o = initial available BOD
ℓ = BOD applied per loading
l_o = BOD applied per unit time
T = time period for intermittent loading
n = number of periods
t = time (for continuous loading)
k = oxidation rate constant

Attempts to verify these equations with data from soil columns in the author's laboratories have been unsuccessful. However, both equations predict an equilibrium concentration of organic matter in the soil that increases as the loading increases, and there are data verifying this. For example, Table 7.3 summarizes some data obtained in lysimeter tests designed to

Table 7.3. Equilibrium soybean oil concentrations in Hanford sandy loam as a function of loading[a]

Hydraulic Loading gal/ac day[b]	Oil Loading lb/ac day[c]	Equilibrium Oil Concentration[d] mg/g soil	Per Cent Decomposition
30,000	12.5	0.10-0.18	92.5
60,000	25.0	0.40-0.50	88.0
80,000	33.5	1.7	71.5
120,000	50.0	2.4	68.0

[a]After Reference 43.
[b]Multiply by 9.35×10^{-3} to convert to m^3/ha day.
[c]Multiply by 0.89 to convert to kg/ha day.
[d]Composite of top one inch (2.54 cm) of soil

determine the degradation rate of soybean oil in sandy soil from the Richland, Washington area. The soybean oil emanates from the oil blanching of potatoes. The oil concentration in the feed to the lysimeters was 50 mg/l; total COD of the feed was 1,000 mg/l.

A rational model of substrate utilization by microorganisms attached to a packed column has been developed recently by Pirt.[44] This model visualizes the packed column (soil column in our case) as a series of "theoretical compartments" each separated by a "theoretical film" of biomass. The liquid passes down the column so that it remains for an equal time in each compartment. The efficiency of the column is represented by the number of theoretical compartments per unit length of column. Pirt developed expressions for substrate consumption in both aerobic and anaerobic columns. For aerobic columns, the assumption was made that oxygen was the rate-limiting nutrient. The functional relationship derived for aerobic columns was

$$S_r - S_e = \frac{nKc^{1/2}}{PF} \quad (3)$$

where

S_r = feed substrate concentration
S_e = effluent substrate concentration
n = number of theoretical compartments in the column
c = dissolved oxygen concentration in the liquid medium
P = oxygen demand constant (grams oxygen consumed/gram substrate oxidized)
F = liquid flow rate
K = $A\,(2D\rho q)^{1/2}$

A = cross-sectional area of column
D = diffusion coefficient for limiting substrate through layer of biomass
ρ = density of biomass
q = specific rate of substrate consumption by biomass (gram substrate/gram dry biomass-hour)

The above relationship has been tested by its originator with limited data and has been found to give reasonable results. More study to develop reasonable values for the parameters in this equation is definitely needed. Pirt's work, probably with some modification to account for both deviations from plug-flow and nonbiological removal mechanisms, seems likely to serve as the starting point for the development of a realistic design approach to land disposal systems.

There are several other variables that could have an effect on the decomposition rate; these bear brief mention as they would need to be measured or controlled in a comprehensive study aimed at producing a rational model. The first of these is temperature. It is well-known that biological processes are temperature-dependent and there is no reason to suspect that decomposition in soil is any exception. Studies by many soil scientists have demonstrated significant correlation between microbial activity as evidenced by carbon dioxide evolution and soil temperature. It may be, however, that the decreases in specific metabolic rate that accompany decreases in temperature are compensated for by increases in the microbial population. This phenomenon was observed by Vela and reported by Gilde, *et al.*[9]

The composition of the inorganic fraction of the waste may have some effect on decomposition rate. Broadfoot and Pierre[45] found a relationship between the decomposition of leaves and their acid-base balance. The decomposition rate was strongly correlated with a property they defined as the excess base: the excess of basic materials (Ca, Mg, Na and K) over acidic materials (Cl_2, S and P).

Finally, it is well established that decomposition of nutrient-poor wastes can be hastened by adding some nitrogen and phosphorus in the form of inorganic salts. Table 7.4 gives the percentage of decomposition of cellulose in a glassine waste applied to soil as a function of the carbon:nitrogen ratio. The data demonstrate convincingly that a nutrient deficiency slows down the decomposition.

Table 7.4. Cellulose decomposition in soil as a function of initial C/N ratio[a]

Time, Days	Per Cent Cellulose Remaining for Various C/N Ratios		
	5:1	10:1	550:1
0	100	100	100
20	80	82	86
30	60	66	76
40	32	50	65
50	0	30	53
60	0	8	38

[a]After Reference 46.

AREAS FOR RESEARCH

The data compiled from the numerous studies that were reviewed for preparation of this chapter indicate that the decomposition rate of industrial wastes applied to the soil depends on the nature of the organic materials and the environmental conditions, including soil properties, under which the decomposition takes place.

The nature of needed research, then, relates to making decomposition rate measurements in systems where there is sufficient control over environmental variables to allow a fairly general application of the data obtained. In practice this will not be as simple as it sounds. The interrelationships between the many variables that play a role are complex enough to defy rapid progress in this area. However, when one reviews the incredible number of research projects funded to determine the "Fate of Waste A in Biological System B," it would seem appropriate to expect some funds to be directed towards fundamental and applied work in soil systems.

In order to control adequately the many variables encountered, the bulk of this work will probably involve the use of laboratory lysimeters. These soil columns could be used to study both hydraulic and organic loading patterns under a wide variety of conditions. They would normally produce data on percolate quality, sustained infiltration rates and changes in soil physical and chemical properties, all of which are critical to the design of a full-scale system. Lysimeter studies are not required preparatory to all, or even most, designs. However, their use is indicated in the case of industrial effluents, high-rate systems, and critical groundwater quality situations.

REFERENCES

1. Anderson, D. J. and A. T. Wallace. "Innovations in Terrestrial Disposal of Steam Peel Potato Wastewater," *Proc. of the Pacific Northwest Industrial Waste Management Conference* (October 1971), p. 45.
2. Sandborn, N. H. "Disposal of Food Processing Wastes by Irrigation," *Sew. Ind. Wastes* **25**, 1034 (1953).
3. Lane, L. C. "Disposal of Liquid and Solid Wastes by Means of Spray Irrigation in the Canning and Dairy Industries," *Proc. of the 10th Industrial Waste Conference*, Purdue Univ. **89**, 508 (1955).
4. Dietz, M. R. and R. C. Frodey. "Cannery Waste Disposal by Gerber Products," *Compost Science* **1**, 22 (1960).
5. Monson, H. "Cannery Waste Disposal by Spray Irrigation—After 10 Years," *Proc. of the 13th Industrial Waste Conference,* Purdue Univ. **96**, 449 (1958).
6. Canham, R. A. "Comminuted Solids Inclusion with Spray Irrigated Canning Waste," *Sew. Ind. Wastes* **30**, 1028 (1958).
7. Luley, H. G. "Spray Irrigation of Vegetable and Fruit Processing Wastes," *J. Water Poll. Control Fed.* **35**, 1252 (1963).
8. Bendixen, T. W., *et al.* "Cannery Waste Treatment by Spray Irrigation-Runoff," *J. Water Poll. Control Fed.* **41**, 385 (1969).
9. Gilde, L. C., *et al.* "A Spray Irrigation System for Treatment of Cannery Wastes," *J. Water Poll. Control Fed.* **43**, 2011 (1971).
10. Hands, F. J., J. R. Lambert, and P. S. Opliger. "Hydrologic and Quality Effects of Disposal of Peach Cannery Waste," *Trans. Amer. Soc. Agric. Eng.* **11**, 90 (1968).

11. Ludwig, R. G. and R. V. Stone. "Disposal Effects of Citrus By-Products Wastes," *Water Sew. Works* **109**, 410 (1962).
12. Anderson, D. R., *et al.* "Percolation of Citrus Wastes Through Soil," *Proc. 21st Industrial Waste Conf.*, Purdue Univ. **121**, 892 (1966).
13. Molloy, D. J. " 'Instant' Waste Treatment," *Water Works Wastes Eng.* **1**, 68 (1964).
14. Breska, G. J., *et al.* "Objectives and Procedures for a Study of Spray Irrigation of Dairy Wastes," *Proc. of the 12th Industrial Waste Conference,* Purdue Univ. **94**, 636 (1957).
15. Lawton, G. W., *et al.* "Spray Irrigation of Dairy Wastes," *Sew. Ind. Wastes* **31**, 923 (1959).
16. McKee, F. J. "Spray Irrigation of Dairy Wastes," *Proc. of the 10th Industrial Waste Conf.*, Purdue Univ. **89**, 514 (1955).
17. Scott, R. H. "Disposal of High Organic Content Wastes on Land," *J. Water Poll. Control Fed.* **34**, 932 (1962).
18. Henry, C. D., *et al.* "Sewage Effluent Disposal Through Crop Irrigation," *Sew. Ind. Wastes* **26**, 123 (1954).
19. Schraufnagel, F. H. "Ridge-and-Furrow Irrigation for Industrial Waste Disposal," *J. Water Poll. Control Fed.* **34**, 1117 (1962).
20. "Pulp and Paper Mill Waste Disposal by Irrigation and Land Application," *Stream Improvement Technical Bulletin* No. 124, National Council of the Paper Industry for Air and Stream Improvement, Inc. (December 1959).
21. "Recent Studies of Irrigation Disposal of Pulp Mill Effluents," *Stream Improvement Technical Bulletin* No. 150, National Council of the Paper Industry for Air and Stream Improvement, Inc. (September 1961).
22. Wisniewski, T. F., A. J. Wiley, and B. F. Lueck. "Ponding and Soil Filtration for Disposal of Spent Sulphite Liquor in Wisconsin," *Proc. of the 10th Industrial Waste Conf.*, Purdue Univ. **89**, 480 (1955).
23. Billings, R. M. "Stream Improvement through Spray Disposal of Sulphite Liquor at the Kimberly-Clark Corporation, Niagara, Wisconsin, Mill," *Proc. of the 13th Industrial Waste Conf.*, Purdue Univ. **96**, 71 (1958).
24. Gellman, I. and R. O. Blosser. "Disposal of Pulp and Papermill Waste by Land Application and Irrigation Use," *Proc. of the 14th Industrial Waste Conf.*, Purdue Univ. **104**, 479 (1959).
25. Blosser, R. O. and E. L. Owens. "Irrigation and Disposal of Pulp Mill Effluents," *Water Sewage Works* **111**, 424 (1964).
26. Crawford, S. C. "Spray Irrigation of Certain Sulfate Pulp Mill Wastes," *Sew. Ind. Wastes* **30**, 1266 (1958).
27. Wallace, A. T., R. Luoma, and M. Olson. "Studies of the Feasibility of a Rapid-Infiltration System for Disposal of Kraft Mill Effluent," to be presented at the 30th Industrial Waste Conf., Purdue Univ. (1975).
28. Voights, D. "Lagooning and Spray-Disposal of Neutral Sulfite Semi-Chemical Pulp Mill Liquors," *Proc. of the 10th Industrial Waste Conf.*, Purdue Univ. **89**, 497 (1955).
29. Meighan, A. D. "Experimental Spray Irrigation of Strawboard Wastes," *Proc. of the 13th Industrial Waste Conf.*, Purdue Univ. **96**, 456 (1958).

30. Parsons, W. C. "Spray Irrigation of Wastes from the Manufacture of Hardboard," *Proc. of the 22nd Industrial Waste Conf.*, Purdue Univ. **129**, 602 (1967).
31. Philipp, A. H. "Disposal of Insulation Board Mill Effluent by Land Irrigation," *J. Water Poll. Control Fed.* **43**, 1749 (1971).
32. Koch, H. C. and D. E. Bloodgood. "Experimental Spray Irrigation of Paperboard Mill Wastes," *Sew. Ind. Wastes* **31**, 827 (1959).
33. Flower, W. A. "Spray Irrigation for the Disposal of Effluents Containing Deinking Wastes," *TAPPI* **52**, 1267 (1969).
34. Parker, R. R. "Disposal of Tannery Wastes," *Proc. of the 22nd Industrial Waste Conf.*, Purdue Univ. **129**, 36 (1967).
35. Colovos, G. C. and N. Tinklenberg. "Land Disposal of Pharmaceutical Manufacturing Wastes," *Biotech. Bioeng.* **4**, 153 (1962).
36. Woodley, R. A. "Spray Irrigation of Organic Chemical Wastes," *Proc. of the 23rd Industrial Waste Conf.*, Purdue Univ. **132**, 251 (1968).
37. Lever, N. A. "Disposal of Nitrogenous Liquid Effluent from Modderfontein Dynamite Factory," *Proc. of the 21st Industrial Waste Conf.*, Purdue Univ. **121**, 902 (1966).
38. Hickerson, R. D. and E. K. McMahon. "Spray Irrigation of Wood Distillation Wastes," *J. Water Poll. Control Fed.* **32**, 55 (1960).
39. Ongerth, J. E. "Feasibility Studies for Land Disposal of a Dilute Oily Wastewater," to be presented at the 30th Industrial Waste Conf., Purdue Univ. (1975).
40. Minderman, G. "Addition, Decomposition and Accumulation of Organic Matter in Forests," *J. Ecol.* **56**, 355 (1968).
41. "Sewage Treatment at Military Installations," National Research Council Subcommittee Report, *Sew. Works J.* **18**, 796 (1946).
42. Costopoulos, J. M. "An Experimental Study on the Rate of Oxidation of Organic Wastes in Soil," unpublished Ph.D. Thesis, Northwestern Univ. (1959).
43. Brown, S. J. "Oil Decomposition in Soil Disposal Systems," unpublished M.S. Thesis, Univ. of Idaho (1972).
44. Pirt, S. J. "A Quantitative Theory of the Action of Microbes Attached to a Packed Column: Relevant to Trickling Filter Effluent Purification and to Microbial Action in Soil," *J. Appl. Chem. Biotechnol.*, **23**, 389 (1973).
45. Broadfoot, W. M. and W. H. Pierre. "Forest Soil Studies: I. Relation of Rate of Decomposition of Tree Leaves to their Acid-Base Balance and Other Chemical Properties," *Soil Science* **48**, 329 (1939).
46. Blosser, R. O. and A. L. Caron. *Technical Bulletin* No. 185, National Council of the Paper Industry for Air and Stream Improvement. (1965), p. 43A.

8

APPLICATION OF AGRICULTURAL WASTES TO LAND

Raymond C. Loehr

Director, Environmental Studies Program,
Professor, Agricultural and Civil Engineering Departments
College of Agriculture and Life Sciences
Cornell University
Ithaca, New York

Stuart D. Klausner, Thomas W. Scott

Research Associate and Professor
Department of Agronomy
College of Agriculture and Life Sciences
Cornell University
Ithaca, New York

INTRODUCTION

The 1972 Amendments to the Federal Water Pollution Control Act note that it is a national goal to minimize the discharge of wastes to surface waters. This goal will result in greater attention to water conservation and reuse, to waste utilization possibilities, and to greater interest in the use of the land as an acceptor of wastes. The Amendments also require the identification of nonpoint sources of potential pollution, including runoff from waste disposal areas, and of feasible procedures to control such sources.

Engineers and scientists have been concerned with the assimilative capacity of streams, estuaries, and lakes for decades while comparatively little is known about the waste assimilative capacity of a soil. Each soil has a maximum capacity to assimilate wastes and to renovate wastewaters. The capacity is related to the soil characteristics, environmental conditions, and the crops grown.

Each disposal site has a controlling parameter that limits the waste assimilative capacity of the site. The controlling parameter is a function of the characteristics of the waste, characteristics of the soil, and items of environmental concern. Examples of such items include salt accumulation, flooding caused by excess applied water, and excess nitrogen and phosphorus in soil water and runoff. Acceptable waste loading rates to the soil are known only in general terms. Detailed information that permits engineered use of the land as an acceptor of wastes is needed. The talents of many disciplines, such as agronomy, agricultural engineering, sanitary engineering, and economics, are needed to develop criteria that will permit acceptable use of the land as a resource to accommodate the residues of man and agriculture.

Much of the information available on feasible land waste loading rates has been obtained from studies on the disposal of animal manures and food processing wastes. The purposes of this chapter are: (1) to present the fundamentals relating to the use of soil for the disposal of wastes and (2) to illustrate the use of the fundamentals by examples of the disposal of manure on land in New York and northeastern United States.

FUNDAMENTALS

General

The land represents not only an appropriate disposal medium for wastes but also an opportunity to manage wastes with minimum adverse environmental effects. Application of manure, sewage sludge, municipal wastewater, and industrial wastes on land for both disposal and fertilizer value has been practiced for centuries. The challenge is to utilize the chemical, physical, and biological properties of the soil as an acceptor for the residues of man with minimum unwanted effects to the crops that are to be grown, to the characteristics of the soil, and to the quality of the groundwater and surface runoff.

The soil system has complex chemical, physical and biological properties which determine the suitability of the soil for growing plants. Other factors such as topography and climate may determine the extent to which the soil can be utilized for plant growth. The soil properties that are important to growing plants determine the fate of waste materials applied to or incorporated in the soil. One must consider the significance of some of these soil properties to have an understanding of the proper use of the soil for waste disposal or, perhaps more properly, waste utilization. Since there is a large array of soil types, a knowledge of some of the important properties will permit a more judicious use of the soil and can be used to establish acceptable loading limits of the soil. Unnecessary pollution of groundwater, streams, and lakes can occur because soil properties are not known or understood.

Until recent times, animal wastes were utilized as a valuable economic source of nutrients for crop production. Since World War II, commercial fertilizers have become the preferred source for supplementing nutrients in the soil because of their relatively low cost, ease of handling, ease of storing and ready availability. It is imperative that ways are found to utilize agricultural wastes to improve soils and provide added fertility for plant growth.

Soils vary greatly in their physical and chemical properties and are classified according to these properties. The soil series is the lowest unit in the soil taxonomy classification scheme. In New York, for example, there are over 300 soil series. It would be a complicated task to evaluate individual soils for waste disposal purposes. By knowing some important soil properties, soils can be grouped into a few categories that can provide a valuable guide for waste disposal purposes. Such information is contained in soil survey reports and can be obtained from the local Soil Conservation Service office.

The soil survey reports show many of the chemical and physical properties of soils. An understanding of these properties provides information needed to determine the suitability of soils for land disposal of wastes. The engineer may find the advice of a soil scientist valuable when attempting to locate soils for waste disposal.

Soil Chemical Properties

Soils of the humid temperate regions have a net negative charge that permits them to retain or hold the positively charged cations, thus providing a reservoir or storehouse for plant nutrients. This charge originates with the very small size fractions of clay and humus in the soil. The extent of this ability to hold cations is termed the cation exchange capacity and is expressed in milliequivalents (me) per 100 grams of soil. Cations held by this exchange complex are essential plant nutrients such as calcium, magnesium, potassium, ammonium, sodium, and aluminum. The source of these cations can be the result of mineral weathering, organic matter decomposition, applied fertilizers, or animal wastes.

Negatively charged anions such as nitrate and chloride are not affected by the temperate region soil charges and move freely with soil water. Nitrates moving beyond the rooting depth of crops can affect water quality. Movement of nitrogen through the soil profile by percolating water is one of the main mechanisms by which nitrates are lost from the system. One of the determining factors in the use of land for waste disposal is the amount of nitrates in the soil and the amount of water moving through the soil. This is not true for the other main fertilizing elements–phosphorus and potassium. Nitrogen and phosphorus are emphasized in this discussion because they are considered major sources of potential water quality problems. They are found in large quantities in animal wastes and are applied in commercial fertilizers.

Soil Reaction

The chemical conditions existing in soils determine the reaction of soil, which may be acid, neutral or alkaline. This reaction in turn determines the availability or solubility of certain elements as well as the response of microorganisms and higher plants.

There is a natural tendency for soils to become acid in humid climates where sufficient rainfall occurs to leach bases from the surface layers. Hydrogen and aluminum become dominant in the exchange complex and the soil is acid in reaction. When soil colloids are dominated by calcium and magnesium on their adsorptive surfaces, the soil is neutral or alkaline in reaction. This condition occurs in limed soils or low rainfall areas. More than one-third of New York agricultural soils test in the acid range or below pH 6.0. Most crops grow best in the slightly acid to neutral pH range of 6.0 to 7.0.

The soil chemical properties determine the capacity of the soil to break down the complex waste materials added in varying amounts. These properties are also influenced by the application of wastes.

Physical Properties

The physical properties of soils determine the rate at which water moves (infiltration rate) into and through (leaching or percolation) the soil. Texture and structure determine these factors. Texture refers to size groupings of individual particles such as sand, silt, and clay. Structure refers to the grouping of the individual particles into aggregates. These factors determine the porosity or pore size distribution in soils, which is important to the hydraulic conductivity of the soil.

There is little phosphorus movement within the soil. Phosphorus entering streams from fertilized crop land results primarily from soil erosion. The movement of nitrates in the soil depends not only on the amount of rainfall and evapotranspiration, but also the water-holding capacity of the soil. Nitrogen movement occurs more readily in sandy soils than in silt and clay soils because the water-holding capacity is greatest for clay soils, less for silts, and least for sands. Harmsen and Kolenbrander[1] noted that the vertical downward displacement of nitrogen in sandy soils, beginning with moisture content around field capacity, was about 18 inches per 4 inches (45 cm/ 100 mm) entering the surface, about 12 inches (30 cm) in soils with 20-40% of the particles less than 200 microns in diameter, and only about 8 inches (20 cm) displacement per 4 inches (100 mm) of rainfall for heavy clay soils. Such data help explain the slow rate of leaching for medium and heavy textured soils in areas of relatively high rainfall. Where leaching losses do occur from soils other than sands, it is more likely to occur during fall or spring months. Little nitrogen loss by leaching occurs during the growing season unless atypical rainfall patterns are prevalent.

Other physical properties of importance when considering land for waste disposal include location of dense subsurface layers, pans, and water tables. Anything that restricts downward water movement and causes oversaturation of the surface soil will encourage surface runoff. Good infiltration of water and sufficient soil aeration are essential for good crop growth and microbial activity for the breakdown of organic material.

Nitrogen

Nitrogen exists in soils in both organic and inorganic forms. Most of the soil nitrogen is associated with organic forms added as crop residues and/or animal wastes. All of this nitrogen originated in the atmosphere and eventually reached the soil through various fixation processes. Relatively small amounts of inorganic nitrogen occur in the soil at any one time. Most of this inorganic nitrogen occurs in the ionic forms of ammonium (NH_4^+), nitrite (NO_2^-) and nitrate (NO_3^-). The inorganic forms are the principal forms of nitrogen used by plants with the exception of plants that fix atmospheric nitrogen in symbiosis with microorganisms.

In unfertilized soils, organic matter is the nitrogen source, and complex soil processes bring about the conversion of this organic nitrogen to the available inorganic forms noted above. The rates of these transformations are dependent upon the forms of nitrogen present, the temperature, and the quantity of active organisms. The formation of ammonium nitrogen can be accomplished by large numbers of heterotrophic organisms while the formation of nitrite and nitrate is accomplished primarily by autotrophic microorganisms that require elemental oxygen and well-aerated soils.

Soil organic matter can provide a suitable source of nitrogen for crops. The conversion of organic nitrogen to inorganic nitrogen occurs at a rate of about 1.2 to 2% per year. The nitrogen that is mineralized is available for crop growth. The nitrogen released from soil containing about 4% organic matter in the plow layer can support corn yields of 50-80 bushels without the addition of fertilizer or manure nitrogen.

All animal manures added to the soil should be part of a crop production scheme. Nitrogen added in manures is mostly organic, since much of the inorganic ammonia may have been volatilized before soil incorporation. Release of nitrogen from animal manure as ammonium occurs under a wide range of soil conditions but is favored by temperatures above 60°F (16°C). Ammonium ions (NH_4^+) can be adsorbed on the exchange sites of the soil and their mobility reduced. However, the ammonium can be nitrified to nitrate (NO_3^-), which is soluble and mobile. Nitrate is easily lost by runoff or by leaching to deeper depths in the soil. In the soil profile, much of the nitrate can be reduced, usually to elemental nitrogen (N_2) if denitrifying conditions exist, and lost to the atmosphere.

Conservatively, the quantity of manure or other wastes added to soils should be such that the mineralized nitrogen will be utilized by a crop during a given season. However, a portion of the mineralized nitrogen may be denitrified and actual manure loading rates can be higher than this conservative estimate. Because the quantity of nitrate nitrogen that is denitrified is difficult to predict, caution should be used in applying manures at rates considerably in excess of the conservative estimate. Quantities of nitrogen added above that which can be utilized by a crop will eventually be lost from the agricultural sector.

Phosphorus

Phosphorus presents less of a pollution hazard than nitrogen in groundwater. While nitrate can move freely with soil water, phosphorus is relatively immobile in the soil. It exists in the soil in both organic and inorganic forms, roughly in equal parts. Organic phosphorus is mineralized in the soil, and applied or mineralized inorganic phosphorus is converted rapidly to water-insoluble forms. The inorganic forms exist as calcium, iron, and aluminum phosphates. The mobility of phosphorus compounds in a soil is low and only small amounts are expected in groundwater. Soil erosion processes, however, may move substantial amounts of phosphorus as constituents of soil particles and soil organic matter.

Water Movement

Surface and subsurface flow of water is the transporting mechanism for pollutants from the land surface. Though a watershed may have a large capacity to store water within the soil, the infiltration rate of the soil may limit the percentage of rainfall that can be stored. Whenever rainfall occurs more rapidly than it can be absorbed by the soil, it moves across the ground surface into drainage ways and depressions. The water flowing over the soil surface carries with it dissolved chemical elements as well as suspended solids. Water not lost by surface runoff infiltrates and percolates through the soil profile. Some of this water may reappear downslope as groundwater seepage or may percolate into the groundwater reservoir, carrying dissolved nutrients to the underground aquifer.

Since a cyclic system for nutrients contained in wastes is important in their land disposal, management practices must center around soil characteristics and the climatological patterns prevalent in a particular area. Animal wastes must remain in place for as long as possible so as to maximize its utilization.

The relationship between stream flow, precipitation, and evapotranspiration for an "average" year in western New York is noted in Figure 8.1. The pattern is similar in other states in the Northeast, although the actual quantities vary between regions and years in response to local climatic variation. Regardless of the specific relationship for a particular region, the basic concepts apply. When evapotranspiration exceeds precipitation, crops are dependent on water in the soil reservoir. During these periods, movement of water to underground aquifers is at a minimum. When precipitation

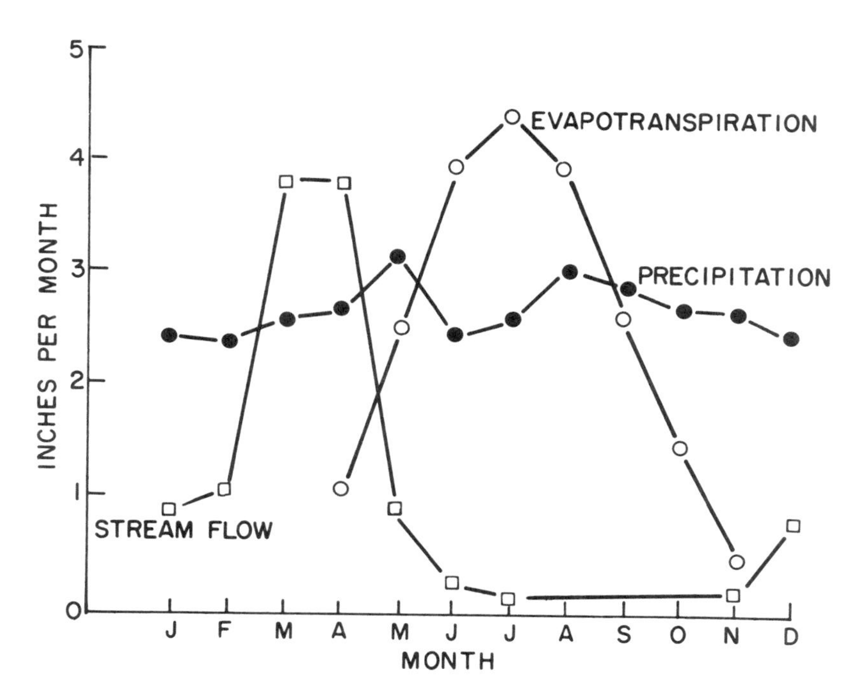

Figure 8.1. Estimated evapotranspiration in the Erie-Niagara Basin, normal rainfall at Lockport, N.Y. and stream flow of Little Tonawanda Creek at Linden, N.Y.[3]

begins to exceed evapotranspiration, there is no appreciable increase in water movement out of the soil profile because of soil water recharge. However, if precipitation continues to exceed evapotranspiration, the capacity of the soil water reservoir may be exceeded. If this occurs, excess water is available for percolation through the soil profile to underground aquifers. Surface runoff, however, can occur any time when the intensity of a rainstorm or snow melt exceeds the infiltration capacity of the soil.

Surface runoff is most prevalent when the soil is near saturation because of the restrictions of water transmission through the profile. This condition correlates closely to periods of groundwater recharge. Excessive movement of water through the soil profile as well as across the surface is reflected in the rise in stream flow. The period when the largest flow of water is likely, and hence, the period of major nutrient transport is during periods of high stream flow.[2] In the northern states, this occurs in late winter and spring runoff. The management of the land application of wastewaters and animal wastes must be oriented around preventing sediment and nutrients from being carried by the runoff and preventing nutrients from being carried below the root zone of the crops by leaching.

EXAMPLES

Manures

Manure can be spread on land either as slurry or as a semi-solid. If the manure has been stored under anaerobic conditions, odorous volatile compounds are released by spreading. An aerated liquid treatment system can be used in conjunction with a slurry waste disposal system to minimize odor problems and facilitate handling of the waste. In addition, odors can be minimized by immediate soil incorporation.

Intermittent waste applications preserve the effectiveness of the soil to assimilate the wastes. For best results, concentrated organic wastes such as manures should be injected directly into the soil or applied to the surface and incorporated with the soil as soon as possible by discing or plowing. The problem of agricultural waste disposal is greater during winter months when lands cannot be plowed and under conditions when the soil is at or near saturation.

Poultry manure and broiler litter have been applied to land under conditions that have caused soil pollution and related problems. Where manure-litter has been returned to the soil at rates exceeding 10 tons per acre annually (t/ac yr), problems of excess salts and a chemical imbalance have occurred in the soil. In addition, problems of nitrate toxicity and tetany have occurred in the grass pasture.[4] As a consequence of these potential health problems, broiler litter disposal rates greater than 4 t/ac yr (9 t/ha yr) are not recommended on fescue pasture systems.[5]

The plow-furrow-cover method was used to compare the disposal of poultry manure at rates of 0 to 45 tons of dry poultry manure per acre (0-100 t/ha).[6] The tests were on a loamy soil with a clay content of 15-33%. The soil did not have a cover crop in these experiments. With time and rainfall, the nitrate nitrogen was distributed in the soil profile. The study concluded that in these soils, poultry manure should be applied at rates less than 15 tons of dry poultry manure per acre (35 t/ha) because of potential nutrient contamination of the soil water.

Manure accumulations on and around beef feedlots can also be a cause for concern. Many management factors affect the amount of nitrate in these soils, such as type of feed, length of livestock feeding, length of lot occupation, and frequency of manure removal. Investigations of the nutrient accumulation under feedlots revealed that soil texture did not affect the nitrate distribution pattern beneath feedlots where the soil was of fairly constant composition.[7] When the soil profile varied, both moisture and nitrate concentration increased as the clay content increased. Soil phosphorus analyses indicated little, if any, movement of phosphorus. The phosphorus was concentrated in the top layer of the soil or manure layer. Slight phosphorus movement was detected in soils with an extremely low cation exchange capacity. Nitrate concentrations up to 78 mg/l as nitrogen were found in groundwaters under the feedlots. The longer the feedlot was in existence, the greater the amount of nitrate in the soil profile.

The application of feedlot runoff to permeable loams and clay loams drastically reduced the permeability of these soils.[8] This change in characteristics can be an advantage in sealing the land in a feedlot runoff lagoon to avoid groundwater infiltration. It can be a detriment when the runoff is applied to cropland for irrigation or recovery of nutrients. The detrimental role of feedlot runoff to cropland depends on the rate and amount of runoff application and the land management methods.

When feedlot manure was applied to grain sorghum, grain yields were depressed when the manure was applied at annual rates of 120 and 240 wet tons per acre (269-538 t/ha) for two years. Early growth depression was observed on plots receiving 30 and 60 wet tons per acre (67-135 t/ha). This depression was attributed to high ammonium and salt concentrations in the seed zone. The depression could be relieved by applying irrigation water prior to seeding to decrease the ammonium and salt concentrations.[8] For most conditions, an appropriate rate of beef cattle feedlot manure disposal on irrigated grain sorghum was suggested at 10 wet tons per acre (22 t/ha) every three years in the Texas High Plains area.

Research is under way at Cornell University to determine the effects of land application of dairy manure on nutrient losses and water quality. The manure, which comes from a free stall dairy operation, contains between 15 and 20% solids. Three different rates of application (15, 45, and 90 tons of wet manure per acre, or 35, 100, and 200 t/ha) and three different times of the year (winter, spring, and summer) for disposal were selected on the basis of crop requirements and seasonal evapotranspiration, precipitation, and stream flow. The soil is a mixture of the Lima-Kendaia series with slopes ranging from 2-4%. Both soils are slightly acid to neutral, medium-textured, formed in calcareous glacial till, and range from somewhat poorly drained to moderately well-drained.

Comparisons of soluble nitrogen and phosphorus losses in surface runoff for a 45 t/ac (100 t/ha) loading rate is shown in Figures 8.2 and 8.3. In these figures, the 45 t/ac (100 t/ha) rate illustrates the effect of disposal under the following conditions: (1) on top of approximately 10 inches (25 cm) of melting snow over frozen soil (winter), (2) immediately before plowing it under (spring), and (3) topdressing on corn when 0-2 inches high (summer). It should be clearly understood that the winter 45 t/ac (100 t/ha) rate was applied under adverse weather conditions. This shock loading at one time during an active thaw period when the soil is frozen is not

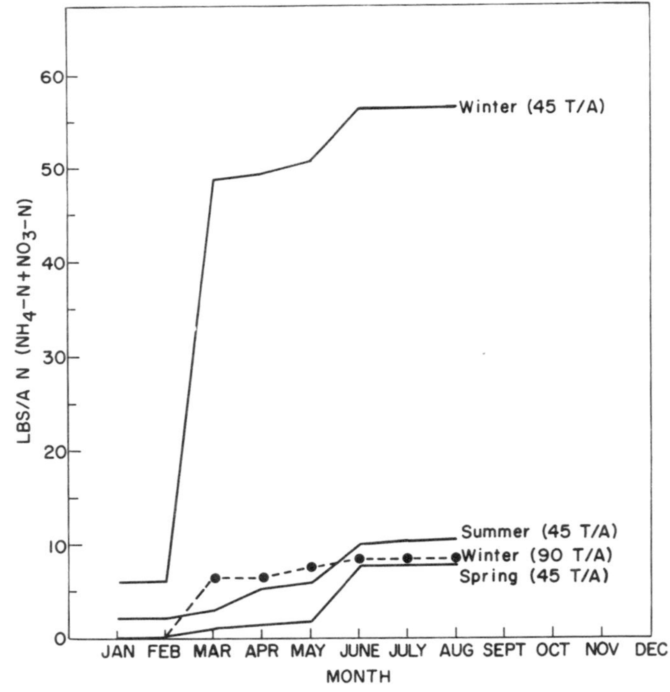

Figure 8.2. Cumulative losses of inorganic nitrogen in surface runoff, 1972.

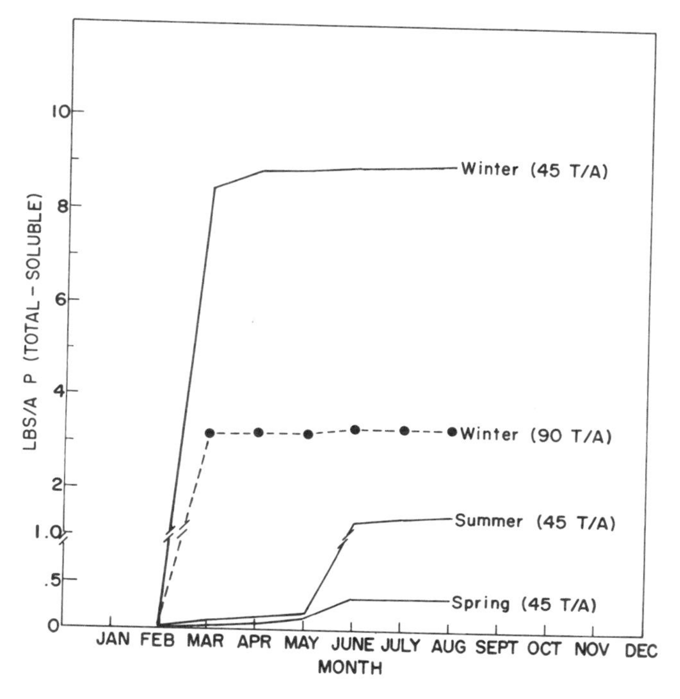

Figure 8.3. Cumulative losses of total soluble phosphorus in surface runoff, 1972.

typical of the array of management practices utilized by Northeast farmers over extensive acreage. In essence, this situation characterizes the extreme.

It is obvious from Figures 8.2 and 8.3 that land disposal of manure during periods of maximum possibility of surface runoff (late winter application) can result in significant runoff losses of soluble nitrogen and phosphorus if the manure is applied on actively melting snow. Approximately 75% and 94% of the nitrogen and phosphorus, respectively, were lost during this time period in relation to the total losses over an eight-month period.

Application of 90 t/ac (200 t/ha) during the winter (Figures 8.2 and 8.3) occurred two weeks prior to a heavy snowfall. The data indicate strongly that winter application during active thaw periods (45 t/ac) can increase nutrient loads drastically. However, winter applications made earlier and then covered by snow, melting at a later date (90 t/ac), showed a much lower level of nutrient losses. The losses of these two nutrients when manure is not applied prior to or during periods of maximum surface runoff is depicted in the relatively lower values during February, March, and April for the spring and summer treatments.

Generally, nutrients in runoff during periods when evapotranspiration exceeds precipitation should be low due to less available water and a greater nutrient uptake by growing crops. The manure during this period was applied on top of emerging corn in mid-June (summer application). Since the summer inputs were not plowed under but exposed on the soil surface, losses of nitrogen and phosphorus from the summer application exceeded those from the spring application. This can be explained by: (1) the Northeast experienced one of the wettest Junes on record during 1972; consequently excessive surface water was present to transport nutrients, and (2) by not plowing under manure, exposure during surface runoff events would result in greater losses.

The data presented are typical of what can happen during a given year. Varying climatic conditions and changing soil conditions cast doubt upon whether these results can be extrapolated to other situations and years. However, the results do indicate the management conditions that can be used to avoid excess nutrient losses when animal wastes are disposed of on the land.

In addition to soil characteristics and climate, vegetative cover on the spreading field needs to be considered. A cyclic system for nutrient application and recovery cannot be maintained if the nutrients are not utilized. The greater the capacity of a given crop to reduce nutrient losses by reducing erosion and utilize the constituents in manure, the greater the amount of manure that can be added. The closer the plant nutrients contained in manure are applied to the time of maximum demand by a crop, the greater the efficiency of utilization. This will, in effect, ensure that the soluble constituents do not remain in the soil for an extended period of time.

In the northeastern states, there are no actively growing crops from early fall until late spring. This period coincides with increased water movement. Consequently, manure additions in the fall and winter create an excess of plant nutrients that are subject to transport. Once these nutrients are either transported off the field or below the root zone, they are essentially lost from the cyclic system, and nutrient enrichment of waters occurs. Conversely, manure applications made during the late spring and summer months, after the potential for maximum runoff and leaching have occurred, allows for more efficient utilization of nutrients by plants. The management practice of disposing of manure on the land and in late spring is not always easy to achieve. In many instances, it is difficult to apply a substantial percentage of the manure just prior to planting a crop, especially if daily spreading is practiced.

Studies at Cornell were conducted on the disposal of treated and untreated poultry manure on soil. The wastes were from a commercial egg production facility and were from an oxidation ditch used to control odors aerobically and to stabilize the wastes.[9] Nitrification occurred in the oxidation ditch, with both nitrites and nitrates being present. When the oxidation ditch mixed liquor was stored in an unaerated tank prior to land disposal, denitrification took place. Oxidation ditch mixed liquor, stored oxidation ditch mixed liquor, and raw poultry manure were used in greenhouse studies to determine their relative effect on the growth of corn.

Raw manure was applied at rates of 0-620 μg N/g of soil. All other sources were applied at rates of 0-1000 μg N/g of soil.[1c] The manure sources contained primarily organic and ammonium nitrogen (Table 8.1).

Table 8.1. Total N and Amount of N in Various Manure Sources[a]

	Stored Oxidation Ditch	Oxidation Ditch	Raw Manure
Total N	0.28	0.32	0.77
Organic N	0.23	0.16	0.25
NH_4-N	0.05	0.04	0.52
NO_2-N	0	0.11	0
NO_3-N	0	0.01	0

[a]Per cent fresh weight basis.

The application of the high rates of manures did not result in excessive amounts of soil soluble salts. The oxidation ditch mixed liquors, stored and unstored, had a slower mineralization rate than did the raw manure. Although each source was applied at equivalent nitrogen loadings, the oxidation ditch mixed liquors had been stabilized prior to land disposal and had a higher percentage of organic nitrogen that was resistant to subsequent microbial degradation. This suggests that higher rates of stabilized aerobically treated animal wastes can be applied to land than can untreated wastes. Similar statements probably are true for treated and untreated municipal, food processing, and other organic wastes.

Soil pH differences affected the rate of nitrogen mineralization from the different manure sources. With an acid soil (pH 4.2), a neutral soil (pH 7.1), and rates of nitrogen up to 1000 μg N/g soil for corn, large differences in mineralization occurred among sources with both soils. In the acid soil, the nitrogen from the stored oxidation ditch mixed liquor was mineralized slower than that from all other sources. Equivalent rates of total N applied to a soil system from different sources supplied different amounts of inorganic nitrogen during a specific time, depending on the rate of mineralization of each source.

Manure sources containing appreciable amounts of NO_2-N, such as the oxidation ditch material, were found to be toxic to plant development only when applied at excessively high rates to neutral soils. This problem can be overcome easily by either storing treated wastes having high nitrite concentrations under conditions favoring denitrification or by applying the material to the soil several weeks prior to planting a crop, thereby allowing ample time for conversion to nitrates. In either situation, the nitrites will be changed to a form that should not inhibit the crops.

Further studies on the disposal of dairy cattle manure and treated and untreated poultry wastes are in progress at Cornell. The continuing studies are large-plot field experiments with loading rates similar to those noted above.

SUMMARY

The application of wastes to land has the potential of causing secondary pollution problems such as salt buildup in the soil, nitrogen and phosphorus in surface runoff, and nitrogen in soil water percolate. With proper waste application rates, land management methods, knowledge of the feasibility of a soil to assimilate wastes, and possible treatment of wastes before application, land disposal of wastes can be an acceptable engineered treatment alternative.

Application to the land is usually the most practical method of manure disposal and utilization. Properly managed manure has distinct value as a source of both plant nutrients and readily available organic matter. When adequate management methods are used, land disposal allows the nutrients from the manure to be recycled to crops to animals to crops again with minimum losses.

A vegetative cover such as grass, legumes, shrubs or trees provides protection against erosion and enhances infiltration. Any means of controlling the movement of water over the surface of the soil (runoff) is helpful in preventing the movement of applied waste. Frequently upland surface flow can be intercepted above the disposal field and delivered to another outlet. A diversion terrace is an effective device for protecting lower lying fields from runoff water originating elsewhere. In the United States, technical assistance with water management problems, erosion control practices, and sites to use for waste disposal are available through the Soil Conservation Service, Soil and Water Conservation Districts and Cooperative Extension Agents.

Manure and similar organic wastes should be incorporated with the soil shortly after spreading, thereby greatly increasing the capacity of the soil to immobilize the components in manure. Incorporation may not be practical except with soil injection systems or applications made prior to spring or fall plowing.

The best time to dispose of manure is when it is most likely to remain where it was applied. The application of manure to land containing the greatest amount of vegetation or crop residue reduces the mobility of the manure. Winter spreading poses numerous problems, the major ones being frozen soil or deep snow that make fields inaccessible. Suitable storage facilities can reduce the problems of winter spreading but may compound problems in the spring.

The appropriate rate of manure application is determined by the ability of the soil-crop combination to immobilize and utilize the applied manure. The greater the crop requirements for plant nutrients, the greater the amount of manure that can be added. Soil characteristics may be the limiting factor, and soil depth, drainage and slope must be considered. The applied nutrients must remain on or in the soil long enough to benefit the crop.

Considerably greater information is necessary to permit the engineered use of the land as an acceptor of wastes. Some pertinent information is reported in this paper and additional studies are underway. Opportunities abound for innovating research and management approaches for using the land.

ACKNOWLEDGMENTS

The research conducted in the College of Agriculture and Life Sciences, a a statutory unit of the State University, at Cornell University, was supported in part by Environmental Protection Agency Project S 800767, "Design Parameters for Animal Waste Treatment Systems."

REFERENCES

1. Harmsen, G. W. and G. J. Kolenbrander. "Soil Inorganic Nitrogen," in *Soil Nitrogen Agronomy*, W. V. Bartholomew and F. E. Clark, Eds. **10**, 43 (1965).
2. Bouldin, D. R., W. S. Reid, and D. J. Lathwell. 'Fertilizer Practices which Minimize Nutrient Losses," Cornell University Conference on Agricultural Waste Management, Cornell University, Ithaca, N.Y. (1971).
3. Harding, W. E. and B. K. Gilbert. "Surface Water in the Erie-Niagara Basin, New York," Basin Planning Report ENB-2, State of New York Water Resources Commission (1968).

4. Hileman, L. H. "Pollution Factors Associated with Excessive Poultry Litter (Manure) Applications in Arkansas," Proceedings Cornell Agricultural Waste Management Conference, Ithaca, N.Y. (1970), p. 41.
5. Wilkinson, S. R., J. A. Stuedemann, D. J. Williams, J. B. Jones, R. N. Dawson, and W. A. Jackson. "Recycling Broiler House Litter on Tall Fescue Pastures at Disposal Rates and Evidence of Beef Low Health Problems," *Livestock Waste Management and Pollution Abatement*, Publ. PROC-271, Amer. Soc. Agric. Engrs. (1971), p. 321.
6. "Poultry Manure Disposal by Plow-Furrow-Cover," Final Report, Grant EC-00254, submitted to the Office of Research and Monitoring, Environmental Protection Agency by the College of Agriculture and Life Science, Rutgers Univ. (1972).
7. Murphy, L. S. and J. W. Gosch. "Nitrate Accumulation in Kansas Ground Water," Project Completion Report Proj. A-016-Kan, Kansas Water Res. Inst. (1970).
8. Stewart, B. A. and A. C. Mathers. "Soil Conditions Under Feedlots and Land Treated with Large Amounts of Animal Wastes," presented at the International Symposium on Identification and Measurement of Environmental Pollutants, Ottawa, Canada, June 1971 .
9. Loehr, R. C., D. F. Anderson, and A. C. Anthonisen. "An Oxidation Ditch for the Handling and Treatment of Poultry Wastes," Proc. International Symposium on Livestock Wastes, ASAE (1971), p. 209.
10. MacMillan, K., T. W. Scott, and T. W. Bateman. "A Study of Corn Response and Soil Nitrogen Transformations Upon Application of Different Rates and Sources of Chicken Manure," *Proc. Agricultural Waste Management Conference*, (Ithaca, New York: Cornell University, 1972), p. 481.

9

POTENTIAL PROBLEMS OF LAND APPLICATION OF DOMESTIC WASTEWATERS

G. Fred Lee

Director, Institute for Environmental Sciences
University of Texas at Dallas
Richardson, Texas 75080

INTRODUCTION

In the past several years, interest in land disposal of domestic wastewaters has increased. This increase arises from a widespread desire to conserve water by recycling. Also, it is thought that land disposal of wastewater would minimize water pollution problems attributed to the presence of large amounts of chemical constituents that can cause significant water quality deterioration in water-based disposal systems. Additional interest in land disposal has been created by the possibility that nutrients present in domestic wastewaters, such as nitrogen and phosphorus, can be recycled to the land where they could then serve as fertilizer for terrestrial plants. Land application of domestic wastewaters is potentially an ecologically sound practice; however, a number of potential problems associated with such a practice could result in environmental degradation.

It is the purpose of this chapter to review some of these potential problems and to discuss an approach for their investigation and evaluation. It will focus primarily on land application systems that involve percolation of the wastewater through several or more feet of the soil column. Little attention will be given to water quality problems associated with overland runoff-type disposal systems.

No attempt will be made in this chapter to provide a comprehensive review of the literature in the topic area. Those wishing additional information on many topics discussed in this chapter should consult the recent reviews by Beatty,[1] Christman,[2] Reed,[3] Thomas,[4,5] and Thomas and Harlin.[6] Several of these discuss the problems normally associated with overland runoff disposal systems.

LAND DISPOSAL VERSUS WATER DISPOSAL

The primary cause of concern about land disposal or application of domestic wastewaters is the presence in these wastewaters of various types of chemical pollutants that could cause water quality deterioration in surface and especially groundwaters around the disposal site. In secondarily treated domestic wastewaters, many chemicals exceed their critical concentrations regarding adverse effects on water quality. However, normally the discharge of domestic wastewaters into surface waters results in a fairly rapid dilution of the chemical constituents to less than critical levels. Thus, dilution is an important mechanism by which potential adverse effects can be minimized. However, rapid dilution is not available with land disposal. This is perhaps the most important fundamental difference between the two basic types of wastewater disposal.

Groundwaters, which in many instances are the ultimate receptacle for domestic wastewaters applied on land, usually mix rather poorly and flow very slowly (typically less than a foot per day). Certain pollutants can be transported long distances in the aquifer without appreciable dilution, and many years may elapse before these contaminants derived from land disposal become significant water quality problems. These problems usually manifest themselves when the contaminated aquifer reaches a domestic or industrial water supply well. Once contaminated, the groundwater aquifer system tends to remain so for long periods of time, even if the source of the problem, *i.e.,* domestic wastewater disposal, is eliminated.

Most professionals in the water supply and pollution control field would place the highest priority on the maintenance of contamination-free groundwater supplies. Therefore, since domestic wastewaters contain a number of chemicals that could readily contaminate groundwaters, any land disposal activities should be accompanied by a detailed groundwater monitoring system, designed to detect any significant contamination before it becomes widespread in the aquifer.

THE "LIVING FILTER" CONCEPT

Some advocates of land disposal of wastewaters postulate that the soil serves as a "living filter," removing "all" chemical contaminants present in domestic wastewaters. One advantage of land disposal is that soil does act as a physicochemical "filter" by removing colloidal and larger particles from the wastewater. However, while such soil-based disposal systems can provide treatment for large quantities of wastewater, they are rarely 100 per cent efficient.

The "living filter" theory is based partially on the biodegradation (and, to some extent, chemical degradation) that occurs in soil-based systems. Further, soil may have a significant capacity to sorb dissolved solutes present in wastewaters. In addition, certain types of chemical transformations, such as precipitation, may occur in these systems. Such reactions would tend to remove chemicals from the wastewater and prevent their entering the aquifer. Finally, it is known that during certain times of the year, terrestrial plants remove some chemical constituents by uptake of these constituents through their root system.

With respect to degradation, the soil-water system can be an effective bio-oxidation system for removal of large amounts of readily degradable organic compounds. Normally, soil percolation removes BOD and suspended solids well. Certain organics, however, are poorly degraded and tend to either accumulate in the surface layers of the soil or permeate the aquifer system. One of the most notorious examples of the latter is the problem that arose from the use of alkyl-benzene sulfonate-type (ABS) detergents. During the early to mid- 1960s the groundwater in various sections of the U.S. was contaminated with ABS as a result of septic tank wastewater disposal systems. In these areas, ABS was poorly degraded in the subsoil system and rapidly transported through the subsoil to the aquifer.

A potentially significant problem associated with large amounts of biodegradation in the soils is the utilization of oxygen present in the waters and production of large amounts of CO_2. Situations could develop where the BOD of the wastewaters would allow little, if any, dissolved oxygen to reach the aquifer, which would have a relatively low pH due to its high CO_2 content. Such conditions could promote the dissolution of iron and manganese in the aquifer and the transport of these chemicals to nearby wells or springs. The low pH would also tend to increase the hardness of the aquifer waters, because of the much greater dissolution of calcium carbonate in hard water areas.

Care must be taken in any process to ensure that the soil filters do not become clogged with the organic solids normally present in domestic wastewaters. Usually, a resting period is necessary whereby the organic solids present can be utilized by bacteria in the soil system. Such resting periods normally increase the rate at which the infiltration of the wastewaters takes place.

The ability of a soil to remove chemical constituents by sorption processes depends on many different factors. One of the overriding factors is the hydraulic loading and interstitial velocity of the water passing over the soil particles. Many types of sorption phenomena are time-dependent, especially those sorption processes involving the migration of the chemical constituents into the pores of the soil particles. It is possible to readily saturate the surface sites of a soil particle with various types of chemical constituents and still have a large potential sorption capacity. However, adequate contact time must be allowed for the uptake of the solute in the capillary pores of the solid.

Another important factor influencing the sorption capacity of aquifer materials and soil systems is the chemical characteristics of the solid phase. Clay minerals, among other types of solids, have a very high sorption capacity for various types of chemical solutes. However, some solid materials, such as quartz or silicate sands with low iron, aluminum, calcium carbonate and organic content, have little or no sorption capacity for most solutes. Normally, based on hydraulic loading, aquifer systems with this type of soil readily receive large amounts of domestic wastewaters. However, such systems have relatively little ability to retain significant amounts of various chemical solutes that can cause water quality problems in the groundwater system.

The transport of phosphate from domestic wastewater septic tank disposal systems is an example of this type of problem. Normally, phosphates are readily held in soil systems. The finite capacity for sorption removal of phosphate can be very high. With precipitation processes, such as the formation of hydroxyapatite, the capacity can be virtually infinite if there is adequate time for precipitation. In quartz-sand soil systems, however, the sorption capacity for phosphate is often low. As a result, groundwaters from these areas may often contain significant amounts of phosphate, which could stimulate algal growth in some surface waters of the region.

One of the methods sometimes proposed to eliminate groundwater contamination by chemicals present in domestic wastewaters is the use of an underdrain or recovery well collection system some six to ten feet below the surface. The purpose of this collection system would be to intercept the wastewaters before they reach the groundwater. While most of the removal of many chemicals generally occurs in the surface of the soils, significant removal for certain chemicals occurs at some depth. It is possible that in time the relatively short soil profile that would be available for removal of chemicals in such systems would result in surface water contamination due to saturation of the soil column.

It is conceptually correct to state that soil systems provide potential for the removal of large amounts of chemical constituents present in domestic wastewaters. However, many factors affect removal; some chemicals are poorly removed in these systems. Further, a particular removal process.

such as biochemical oxidation, may change the characteristics of the percolation water in such a way that it may become contaminated in the aquifer due to the dissolution of aquifer materials.

POTENTIAL CHEMICAL PROBLEMS FROM LAND APPLICATION

Some of the potential chemical problems arising from land application of domestic wastewaters are discussed below.

Nitrate

Most domestic wastewaters contain relatively large amounts of organic nitrogen and ammonia. In an aerobic system, the reduced forms of nitrogen are oxidized to nitrate. Nitrate, a chemical species that is poorly sorbed by aquifer materials, is readily transported in most groundwater systems. Therefore, if domestic wastewater contains organic nitrogen or ammonia, aerobic degradation processes in the soil would bring about solubilization and hydrolysis of the organic nitrogen to ammonia and the nitrification of the ammonia to nitrite and nitrate. These are aerobic processes, and most of the important microorganisms that cause this transformation require the presence of dissolved oxygen. While ammonia and many forms of organic nitrogen are readily sorbed by aquifer materials and thereby removed from the soil surface layers, this sorption does not prevent the biochemical transformation of the ammonia to nitrate.

However, there are two types of biochemical transformations that tend to reduce the nitrate content of wastewaters percolating through the soil system. One of these is the biological uptake of nitrate by terrestrial plants. While terrestrial plants can remove large amounts of nitrate, often the rates of hydraulic loading from domestic wastewaters are such that nitrate removal is relatively poor, even in intensively developed agricultural areas. Another important factor affecting removal of nitrate by terrestrial plants is the relatively short period each year, *i.e.,* the active growing season, during which this removal can occur. Where the growing season is relatively short, either very large storage facilities would have to be provided to store the wastewaters during periods when nitrate uptake would be poor, or nitrogen would have to be removed from the wastewaters before disposal on land.

A second mechanism by which nitrate may be removed from groundwater systems is biochemical denitrification. Denitrification occurs when nitrate is reduced to nitrogen gas in low dissolved oxygen concentrations. Under these conditions, however, the problems associated with dissolution from the aquifer material of iron, manganese, and hardness (due to the low redox potential and pH) will likely occur.

Ten mg/l nitrate-nitrogen is often accepted as the criterion for determining whether excessive nitrate-nitrogen is present in the groundwater system. This value is the normally accepted standard for domestic drinking water. However, since groundwaters frequently come to the surface and become part of the surface water system, and since the algal populations of many surface waters in the U.S. are limited by the inorganic nitrogen content of the water, it is important to consider what critical concentrations of algal nitrogen available in groundwater systems could cause excessive algal growth or eutrophication in surface waters. Generally, it is accepted that whenever the concentrations of inorganic nitrogen as NH_3 and NO_3^--N exceed a few tenths of a mg/l as N, there could be potentially excessive growths of aquatic plants in surface waters.

Chloride

Normally, domestic wastewaters show a dissolved solids content several hundred milligrams per liter over that of the drinking water, and the bulk of this increase is comprised of sodium and chloride. In some hard water areas where ion exchange home water softeners are widely used, the increase in sodium and chloride is markedly greater. Such is the case in Madison, Wisconsin, where the chloride content of the domestic drinking water is in the order of a few milligrams per liter; domestic wastewaters in Madison have chloride contents of 400 to over 500 mg/l. A significant part of this increase is due to the use of home water softeners, although meat-packing operations within the city also account for some of the increase.

The evapotranspiration of the water as it percolates through the soil also tends to increase the dissolved salt content of domestic wastewaters spread on land. This is a significant problem in some of the more arid parts of the United States. The net effect is that the domestic wastewater that percolates through the aquifer may contain chlorides greatly in excess of the drinking water standard of 250 mg/l. Further, these waters could have relatively high concentrations of sodium, which could have an adverse effect on drinking water quality.

Wastewater with high sodium content could also have a significant effect on the sodium adsorption ratio of the water. High sodium waters tend to cause certain clay minerals to swell, thereby inhibiting the ability of the soil to accept large amounts of water. Due to the relatively high cost of salt removal, the high salt content of domestic wastewaters such as those in Madison, Wisconsin, would greatly inhibit the utilization of land disposal.

The problem with excessive salts demonstrates the fundamental difference mentioned earlier between land and water disposal. The rapid mixing that occurs in most surface waters receiving domestic wastewaters normally dilutes these salts below the critical levels. However, in some parts of the U.S., such as the Lake Erie area, concern is focused on the build-up of salts arising in part from domestic wastewater discharge to the lake. Yet, most of these salts, such as sodium and chloride, are very poorly removed in the soil system, and if disposed of on land, would be easily transported through the groundwater. The groundwater system, having relatively poor mixing, would tend to move as a lens or layer that could readily contaminate wells at considerable distances from the point of disposal.

Phosphates

Phosphate concentrations in surface waters are of importance since, for large parts of the U.S., phosphorus is one of the key limiting elements controlling the excessive growth of aquatic plants. Sometimes those working on domestic wastewater disposal systems involving land application tend to overlook the fact that the critical concentrations of phosphate in surface waters are generally less than 10 μg/l. From an agricultural point of view, such concentrations are considered inadequate for growth of many terrestrial plants; however, concentrations of this magnitude readily produce excessive aquatic plant growth. Therefore, in studying phosphate transport through groundwater systems, it is important to measure the phosphorus concentrations down to a few micrograms per liter. Normally, phosphate is held in soil and aquifer systems as a result of precipitation with carbonates and precipitation of and sorption by iron and aluminum oxides and clay minerals. Usually, quartz and other silica and aquifers have very low sorption capacity for phosphate.

Exotic or Hazardous Chemicals

Individuals involved in water pollution control are beginning to realize that domestic wastewaters contain large concentrations of exotic and potentially hazardous chemicals in the wastewater, especially the sludge. Concern has grown over trace metals, some of the most important of which are chromium, cadmium, mercury, zinc and copper. With respect to organic compounds, a wide variety of chlorinated hydrocarbons (*e.g.*, PCB's and phthalates) is present in domestic wastewaters, which could have a significant adverse effect on water quality.

Advocates of land disposal of domestic wastewaters often cite the fact that such practices have gone on for many years in parts of the U.S. as well as elsewhere. They claim that no problems have arisen from the presence of these so-called exotic chemicals in the wastewaters. However, careful examination of the technical literature concerning studies on the effects of land disposal on environmental quality generally shows that they were conducted in such a manner as not to detect problems from these chemicals.

Within the past few years, the generally accepted critical concentrations of many of these chemicals have been drastically reduced. While today many water quality standards are based on acute toxicity to aquatic life, the chronic sublethal toxicity is generally a factor of 10 to 500 times less than the acute lethal toxicity. The chronic sublethal toxicity is generally manifested in reduced growth rates, inhibited reproduction, and other impairments of normal organism activity. This means that, for many chemicals in wastewaters, concentrations considered "safe" in the past will no longer be considered so.

At present, very little information is available on the transport of many of these chemicals in groundwater systems and the translocation of these chemicals from the soil to terrestrial plants. The latter may provide an important mechanism by which some highly toxic chemicals, such as mercury, may be brought back to man through animal feeds grown in domestic wastewater irrigation systems. This problem could become particularly significant if the wastewater disposal system also involves placing significant amounts of the sludge from the wastewater treatment plant on the land.

An additional consideration concerning domestic wastewater land disposal systems is the probability that under certain conditions there will be surface runoff from the disposal area to nearby lakes and streams. This may be particularly important in areas where the soil is frozen, and the spring thaw creates a large overland flow of water that carries with it particles and solutes from previous land disposal of wastewaters. In some parts of the Midwest, such as central Wisconsin, the practice of spreading dairy-barn manure on the frozen soil results in significant transport of phosphorus from the manure via surface flow at the spring thaw. Similar phenomena could occur with some of the chemicals present in domestic wastewaters.

There is an urgent need for detailed studies on the transport of various types of potentially significant chemicals in soil-water and groundwater systems. These studies should also investigate surface water transport of these chemicals from land wastewater disposal areas. It is important that these studies focus on concentrations of these chemical compounds, which are at least a factor of ten less than their currently established chronic sublethal toxicity levels. This means that, in general, the concentrations have to be determined at microgram or submicrogram per liter levels in order to properly define potentially significant transport of these chemicals in the aquatic environment.

OTHER PROBLEMS

While this chapter has focused primarily on chemical problems associated with land disposal of domestic wastewaters, there are certain physical problems that should be recognized, since they could lead to surface water contamination. One of the more significant problems of this type is flooding associated with dike failure. In Madison, Wisconsin, several years ago, a failure of the dikes surrounding a sludge lagoon area resulted in a massive fish kill. The supernatant liquid from the lagoon was transported by heavy rains to nearby streams where its high concentrations of ammonia probably killed large numbers of fish.

In some areas, in order to minimize groundwater contamination problems, it would be necessary to install rather large lagoons for holding wastewaters during certain periods of the year. Unless these lagoons are sealed properly, their increased head would greatly increase the potential for groundwater contamination. Further, periods of heavy rainfall could cause flooding. This would result in large-scale spread of the lagoon waters to nearby surface water courses. It is important that the storage lagoons and the disposal areas be constructed in such a manner as to prevent groundwater contamination and dike failure.

Another problem discussed above and of significance in land wastewater application is the surface runoff associated with winter and spring high flow periods. One problem frequently encountered with wastewater storage lagoons in the Midwest is that during the winter they develop an ice cover, which allows them to go anaerobic. At the spring thaw period, with the loss of the ice, there is often a significant odor problem associated with the release of hydrogen sulfide.

There are also several biological problems associated with land wastewater application that should be mentioned. Depending on the method of land application, there could be increases in insects, such as mosquitoes, as well as rodents. Further, continuous surveillance would have to be provided to prevent possible contamination of surface and groundwaters by human pathogens, although human pathogens present in domestic wastewaters are rarely transported for large distances in groundwaters. The only exception to this would possibly be in some fissured limestone areas. However, some domestic wastewater disposal systems use a spray irrigation method of applying the wastewaters and there is concern that airborne bacteria and viruses may arise from this practice. Only additional study can properly evaluate the significance of this practice to the spread of pathogens to humans.

MINIMIZATION OF WATER QUALITY PROBLEMS FROM LAND WASTEWATER APPLICATION

There is little doubt that in the foreseeable future various parts of the U.S. will resort to much greater utilization of land wastewater application. However, potential problems associated with this practice necessitate careful preevaluation of whether or not these problems will be manifested at any particular location selected for disposal. A discussion is presented below on the topics that should be considered as part of any such evaluation.

Type of Disposal System to be Used

This chapter has focused primarily on a wastewater application system involving groundwater infiltration. However, the type of land wastewater disposal that would be selected for a given area would depend on soil type, land availability, and other factors discussed by Thomas and Harlin.[6] They discuss the three land application systems most frequently used: the infiltration-percolation system, the cropland irrigation system, and the overland runoff system. The first two types of systems have already been discussed.

The third system, overland runoff, is normally used in tight, impervious soils where the wastewaters are spread on the surface, and terrestrial plants provide treatment by removing various constituents from the wastewaters. The waters run off from the disposal area to nearby water courses. Such a system, although it would minimize the problems of groundwater contamination, would have many of the surface contamination problems normally associated with domestic wastewater disposal in waters.

Surface and Groundwater Hydrology

One of the most important factors influencing wastewater application on land is the surface and groundwater hydrology of the disposal area. There is no point in providing a system for groundwater recharge when there is little or no need for groundwater recharge in the region. Similarly, if the groundwater table is near the surface, there is a possibility that rates of infiltration may be poor during the wet part of the year. The various factors important in determining the surface and groundwater hydrology of land wastewater disposal systems are discussed in other chapters.

The importance of surface and groundwater hydrology cannot be overemphasized with respect to the groundwater contamination problems discussed in this chapter. For example, if the groundwater hydrology is such that land disposal will cause the wastewaters to pass rapidly via the aquifer to nearby lakes and streams, then critical concentrations of nitrogen and phosphorus arising from land disposal would have to be utilized in establishing disposal criteria.

Surface and Groundwater Quality

The quality of surface and groundwaters in the region can play an important role in determining the type of treatment and, in some instances, the overall feasibility of wastewater systems utilizing land disposal. If the groundwaters already contain large amounts of certain chemicals, then pretreatment of the wastewaters to remove these chemicals may be necessary in order to keep the total concentrations below the critical levels.

Because of the variable nature of surface water quality, measurements should be made for at least a year to establish the chemical and biological characteristics of the waters prior to any land disposal. It is important to make measurements a function of depth in the aquifer in order to establish whether there is significant stratification occurring prior to any disposal practice.

Soil Aquifer Characteristics for Removal of Contaminants

Some laboratory sorption and leaching studies should be conducted with typical soil and aquifer materials obtained from the proposed application area and with various types of contaminants that are likely to be of significance in the domestic wastewater. Attempts should be made to predict the amount and rates of removal of contaminants in the soil system. If the system is put into operation, this information can then be compared to actual removal. Such information is needed in order to design better laboratory studies that predict the transport of various types of chemical contaminants in soil and groundwater systems.

Wastewater Chemical Characteristics

A one-year period should be used to evaluate the chemical characteristics of the wastewaters to be disposed of on land, with fairly detailed chemical monitoring at approximately bi-weekly to monthly intervals. This monitoring should particularly note certain times of the week, such as late Friday afternoons, when industrial clean-up operations could significantly increase concentrations of certain potentially hazardous chemicals in the domestic wastewaters.

In general, attempts to practice land application of domestic wastewaters that have received little or no pretreatment will likely encounter the greatest difficulties because pretreatment minimizes problems caused by many chemical contaminants. However, this pretreatment may have to be more extensive than conventional secondary treatment for some contaminants. In some instances, more than tertiary treatment might be needed for removal of nitrogen. If the problem is excessive salts, some of the saline water treatment methods may have to be used as pretreatment.

Preliminary Evaluation

From the information obtained, it should be possible to make a preliminary assessment of whether the proposed type of land disposal is acceptable in the proposed area. Specific questions such as, "Will the system work from a technical point of view? Will it cause significant environmental problems?" and, "What will it cost?" must be properly evaluated before any large-scale land application system is put into practice. After such an evaluation, if land application appears to be technically and ecologically sound as well as economically feasible, a pilot-scale project of at least a one year's duration should be used to answer many of the questions about the wastewater application system that are site-specific. For example, at present, it is impossible to predict with a high degree of reliability the rate of transport of various types of chemical contaminants in groundwater systems. However, this prediction should be attempted, and then a pilot-scale study utilizing part of the proposed disposal area should be conducted. In order to evaluate the actual transport of chemical contaminants, the study should utilize anticipated typical loading rates for wastewaters treated in the manner expected. A series of monitoring wells would need to be established in the area. These could be sampled at frequent intervals at various depths in order to determine which of the chemical constituents present in the wastewaters are actually transported and to what extent.

Finally, if the pilot-scale study proves that the system is ecologically and technically sound as well as economically feasible, then the design and operation of the full-scale project could commence. It is essential that the full-scale project have as a key component an adequate groundwater monitoring system to detect long-term effects of land wastewater application. At least at quarterly intervals, samples from various monitoring wells should be collected and examined for the various chemical constituents of interest. Detailed reports on the transport of chemical constituents from the wastewater application area should be prepared annually during the first few years and possibly biennially thereafter. It is also essential that some water quality regulatory authority be assigned the responsibility of examining the groundwater records at periodic intervals in order to detect any minor changes that might signal long-range effects.

CONCLUSIONS

Domestic wastewater disposal on land has been practiced for several hundred years in various parts of the world. It is likely that such practices will receive increased utilization in the U.S. as the need for groundwater recharge increases. It is felt that advocates of land wastewater application systems used in the past have not paid proper attention to the potential problems associated with the chemical contamination of surface and groundwaters in the region of the discharge. Because of the long time often required to remove pollutants from groundwater systems, it is recommended that a conservative approach be used with respect to land application of wastewaters. This approach would involve careful evaluation of all potentially significant chemicals present in the wastewater with respect to their transport in the groundwater system. A long-term ground and surface water monitoring program should accompany any land wastewater application system in order to detect, as soon as possible, any environmental contamination problems arising from the utilization of the system.

ACKNOWLEDGMENTS

This investigation was supported by a training grant from the Environmental Protection Agency. In addition, support was given by the Department of Civil and Environmental Engineering at the University of Wisconsin and the Institute for Environmental Sciences at the University of Texas at Dallas.

REFERENCES

1. Beatty, M. T. *A Technical Evaluation of Land Disposal of Wastewaters and the Needs for Planning and Monitoring Water Resources in Dane County, Wisconsin,* report of the Water Resources Task Group to the Dane County Regional Planning Commission, Madison, Wisconsin (December 1971).
2. Christman, R. F. *Assessment of the Effectiveness and Effects of Land Disposal Methodologies of Wastewater Management,* Wastewater Management Report 72-1, Department of the Army Corps of Engineers (January 1972).
3. Reed, S. C. *Wastewater Management by Disposal on the Land,* Special Report 171, Cold Regions Research and Engineering Laboratory, U.S. Army Corps of Engineers, Hanover, New Hampshire (May 1972).
4. Thomas, R. E. "Applying Wastewaters to the Land for Treatment," *Proceedings of 23rd Industrial Waste and Advanced Water Conference* (Stillwater, Ok.: Oklahoma State University, 1972), pp. 244-252.
5. Thomas, R. E. "Land Disposal II: An Overview of Treatment Methods," *J. Water Poll. Control Fed.* **45**, 1476 (1973).

6. Thomas, R. E. and C. C. Harlin, Jr. "Experiences with Land Spreading of Municipal Effluents," *Proceedings of First Annual Workshop Land Renovation of Wastewater in Florida*, Tampa, 1972 , pp. 151-164.

10

LAND TREATMENT OF WASTEWATER BY INFILTRATION-PERCOLATION

Ronald W. Crites

Project Engineer
Metcalf and Eddy, Inc.
1029 Corporation Way
Palo Alto, California 94303

INTRODUCTION

Infiltration-percolation of wastewater, also known as rapid or high-rate infiltration, is a process that uses the soil matrix for physical, chemical, and biological treatment. The applied wastewater infiltrates the surface and percolates through the soil. Physical straining and filtering occur at the soil surface and within the soil matrix. Chemical precipitation, ion exchange, and adsorption occur as the water percolates through the soil. Biological oxidation, incorporation, and reduction occur principally within the top few feet of soil.

As reported by Pound and Crites,[1] both municipal and industrial wastewaters can be used in the infiltration-percolation process. Feasibility of municipal wastewater treatment by means of infiltration-percolation has been demonstrated by systems operating at Hemet, California,[2] Phoenix (Flushing Meadows), Arizona,[3] Los Angeles (Whittier Narrows), California,[4] and Santee, California.[5] These systems, which have been carefully studied, represent examples of planned land treatment and recovery of the water. Other municipal systems, such as those at Lake George, New York,[6] Westby, Wisconsin,[7] and Marysville, California,[1] represent examples of effluent disposal on the land by infiltration-percolation. Infiltration-percolation systems involving industrial wastewater have been reported at Seabrook Farms, New Jersey,[1] Sumter, South Carolina,[1] and Modesto, California.[8]

The purpose of this discussion is to identify and provide criteria for the major factors relating to design and operation of land application systems using infiltration-percolation. The major topics are (1) site selection criteria, (2) system design, (3) management and operation, and (4) costs.

SITE SELECTION CRITERIA

Criteria for infiltration-percolation site selection are presented for soil characteristics, hydrologic conditions, geologic conditions, and topography. Methods of determining these conditions and more detailed discussions of geohydrology are contained in other chapters.

Soil Characteristics

Well-drained soil, which is essential for infiltration-percolation, includes sand, sandy loams, loamy sands, and gravel. Very coarse sand and gravel are not ideal because wastewater passes too rapidly through the first few feet of soil where the major biological action takes place.[9] These soils are also characterized by low ion exchange capacities and therefore are limited in their chemical retention of pollutants.

A depth of 10 ft (3.05 m) of uniform soil is preferred, although lesser depths can be used if sufficient lateral movement is anticipated. Bouwer[9] indicates that several hundred feet (about 100 m) of underground movement may be required for renovation depending on the effluent quality, renovation requirements, application rate, and soil and subsoil materials.

The depth of soil layers and the permeability of each layer will affect the overall percolation rate. The system may be limited by the presence of a shallow clay layer between two sandy layers because the high permeability of the lower layer will be effectively negated by the presence of such a clay lens. Soil borings,which must be taken at a site to obtain the soil profile, will enable rejection of sites with extensive clay layers or other impermeable formations near the surface.

Hydrologic Conditions

The depth, movement, and quality of groundwater are the most important hydrologic factors in site selection. Wastewater percolating through soil will create a mound that will eventually reach the surface and reduce the percolation rate if there is an insufficient depth to the existing groundwater table or if underdrainage is not provided. Thomas[10] recommends a depth of 15 ft (4.55 m) for maintenance of long-term liquid loading rates and effective renovation. Depths of 10 ft (3.05 m) and less may be used if underdrainage is provided.

The ability of a soil to transmit groundwater determines its suitability to act as an aquifer, which is a geologic formation that contains water and transmits it from one point to another in quantities sufficient to permit economic development. In contrast, an aquiclude is a formation that contains water but cannot transmit it rapidly enough to furnish a significant supply to a well or spring.

A soil with high porosity (the ratio of pore or void volume to the total volume) does not necessarily make a good aquifer. The more accurate measure of the potential of a soil type is the specific yield, which is the ratio of water that will drain freely from the materials to the total volume of the materials. Soil types and their hydraulic characteristics as aquifers are given in Table 10.1. Most of the productive aquifers in the United States are of the sand and gravel type.

Table 10.1. Soil Types and Hydraulic Characteristics[11]

Material	Porosity, %	Specific Yield, %	Permeability, gpd/sq ft[a]
Clay	45	3	1
Sand	35	25	800
Gravel	25	22	5000
Gravel and sand	20	16	2000
Sandstone	15	8	700
Dense limestone	5	2	1
Granite	1	0.5	0.1

[a]The permeability shown is the discharge in gallons per day through an area of 1 sq ft (0.093 m^2) under a gradient of 1 ft/ft at 60°F (15.6°C). Note: 1 gpd/sq ft = 0.283 m^3/min ha

Minimization of the spread of renovated water into an outside aquifer can be accomplished by surrounding the system with wells or drains to interrupt the flow. Evaluating the hydraulics of groundwater flow is an area of increased computer use and modeling. At Flushing Meadows, Arizona, the horizontal and vertical conductivities were calculated by use of an electrical resistance network analog.[3]

The quality and use of the groundwater is critical in site selection. If the groundwater quality meets potable drinking water standards or if the groundwater is being used as a source of drinking water supply, great care must be taken in the design of an infiltration system that directly recharges the aquifer. Conversely, if the existing groundwater quality is poor, the renovated water may be recovered for reuse before it is degraded by the native groundwater. An example of this is in Phoenix where the native groundwater is three times as high in dissolved solids as the secondary effluent applied.[3]

Geologic Conditions

The parent material of a soil is the original material, usually erodible rock such as soft sedimentary rock, from which the soil was created. The parent material gives the soil its chemical and mechanical properties, and knowledge of this parent material and existing rock formations is important in assessing the water-holding characteristics and transmissibility of a soil system.

The geologic formations should be considered in terms of the structure of the bedrock, depth to bedrock, lithology, degree of weathering, and presence of any special conditions, such as glacial deposits. Any discontinuities, such as sink holes, fractures or faults, that may provide short circuits to the groundwater should be noted and thoroughly investigated. In addition, an evaluation of the potential of the area for earthquakes and their probable severity is important in the design of some systems. In most situations, an evaluation by a geologist or geohydrologist is necessary.

Topography

The site topography must be adapted to infiltration-percolation site requirements. Since high-rate infiltration is desired, ponding or flooding the basin is the usual mode of application. A site should be flat or gently sloping so that it can be diked into basins. Too much slope creates lateral percolation, which could reduce the effective percolation rates of the lower basins. Steep slopes also require large quantities of earthwork for basin construction. In such cases sprinkling or contour furrow applications should be investigated.

SYSTEM DESIGN

Infiltration-percolation systems can be designed to meet many different objectives including (1) groundwater recharge, (2) treatment and recovery of water using wells or underdrains, (3) treatment with underflow interception by a surface watercourse, or (4) treatment and disposal of industrial wastewater. The major design factors are:

- preapplication treatment
- liquid loading rate
- organic loading rate
- nutrient loading rate
- basin characteristics
- distribution system
- control of underground flow

Preapplication Treatment

The degree of preapplication treatment necessary depends on the method of application, quality requirements for renovated water, loading rates and cycles, and wastewater quality. Long-term loading rates in flooded basins can be maintained if the suspended solids content is reduced. As an alternative, the wastewater can be applied by large nozzle sprinklers. This method, used on screened food processing wastewater at Seabrook Farms, allows suspended solids concentrations of several hundred milligrams per liter.

For municipal wastewater, preapplication treatment by sedimentation or biological oxidation is generally provided to reduce suspended solids concentrations, to avoid odors, and to increase ease of handling. Disinfection is not generally required as the soil system is quite efficient in removing microorganisms[12,13] Bacteria and viruses and other constituents in renovated water are discussed in the section on Management and Operation.

Hydraulic Loading Rates

Hydraulic loading rates can range from 18 to 500 ft/yr (5.5 to 153 m/yr), or 4 to 120 in./wk (10.2 to 305 cm/wk). The loading is controlled by the saturated vertical permeability and the horizontal flow in the aquifer. Where the horizontal permeability exceeds the vertical permeability, lateral flow will predominate. In this case the hydraulic gradient, the perimeter of the site through which the flow passes, and the boundary conditions will determine the loading rate. As described in Chapter 11, the percolation rate can be estimated through field testing. Most methods, however, produce percolation rates that cannot be attained under full-scale design. Consequently, design hydraulic loading rates should be selected considering field test rates, limiting layer permeabilities, the hydraulic gradient, and groundwater levels.

The long-term hydraulic loading rate for wastewater will be less than the measured rate using the infiltrometer if clear water is used because (1) suspended solids and biological activity clog the surface of the soil, and (2) application periods must be alternated with drying periods to allow the clogging material on the surface to dry and decompose for infiltration recovery. Some existing hydraulic loading rates along with the soil types are given in Table 10.2. Soil permeabilities have been reported for only a few systems, and it appears that they are at least four or five times greater than the average hydraulic loading rate under operating conditions.

Table 10.2. Existing Hydraulic Loading Rates and Soil Types[1,2]

Location	Hydraulic Loading Rate in./wk[a]	Soil Type
Whittier Narrows, California	120	sand
Flushing Meadows, Arizona	72	sand
Santee, California	62	gravel
Lake George, New York	30	sand
Hemet, California	25	sand
Westby, Wisconsin	8.4	silt loam

[a]1 in./wk = 2.54 cm/wk

Organic Loading Rates

Organic loading rates are important in relation to the oxygen diffusion rate and the development of anaerobic conditions. As the soil moisture level increases to values above about 85%, oxygen diffusion into the soil practically ceases. Consequently, to meet the oxygen demand created by the decomposing organic and nitrogenous material, an intermittent loading schedule is required. During the drying period air penetrates the soil and supplies oxygen to the soil bacteria that oxidize the organic matter and ammonium.

Acceptable organic loading rates depend on the soil, the drying or resting period, and the temperature. Reported rates in pounds of BOD_5 per acre per day are given for several systems in Table 10.3. The ratio of drying time to application time is also shown. For the industrial systems, the drying time must be several times longer than the application time, depending upon the soil and climate. For the municipal systems, however, the ratio can be reduced to approximately 1:1. The hydraulic loading cycle, which will be discussed in more detail later, should be selected on the basis of the management objectives as well as the organic loading rate.

The limits of organic loading appear to be on the order of 2,000 lb/ac day (2,242 kg/ha day). Several industrial systems have failed at such loadings. Bogedain, *et al.*[14] report that a failure occurred with milk product wastewater at a loading rate of 1,750 lb/ac day (1,962 kg/ha day). Lawton, *et al.*[15] report loadings up to 1,800 lb/ac day (2,018 kg/ha day) with food processing wastewater and up to 680 lb/ac day (762 kg/ha day) with dairy wastewater. Other organic loadings reported for food processing wastewater range from 400 to 860 lb/ac day (448 to 964 kg/ha day).[16,17]

Table 10.3. Existing Organic Loading Rates and Ratios of Drying Time to Application Time[1,14]

Location	BOD_5 Loading Rate, lb/ac day[a]	Ratio of Drying Time to Application Time
Food Processing Wastewater		
Leicester, New York	500	5:1
Delhi, New York	240	3:1
Sumter, South Carolina	110	2:1
Municipal Wastewater		
Santee, California	57	–
Flushing Meadows, Arizona	45	1:1
Whittier Narrows, California	20	1.6:1
Lake George, New York	19	13:1
Westby, Wisconsin	9	1:1

[a]1 lb/ac day = 1.12 kg/ha/day

For domestic wastewater, Thomas[18] reports that septic tank effluent was loaded successfully on a sandy soil at a rate of 166 lb/ac day (186 kg/ha day). He reports that this loading can be used on sandy soils for extended periods without detrimental accumulation of organic residues in the soil.

Nutrient Loading Rates

The major nutrients of concern–nitrogen and phosphorus–may affect the design of infiltration-percolation systems by reducing the application rates or the long-term use of the site. Other constituents for which reductions are required in the soil system should also be analyzed.[19]

Nitrogen

The primary mechanism for nitrogen removal in infiltration systems is denitrification.[20] The tremendous quantities of nitrogen applied, as much as 21,000 lb/ac yr (23,500 kg/ha yr), far exceed nitrogen-removal capabilities of crops, which may be as large as 600 lb/ac yr (673 kg/ha yr). On the other hand, these tremendous quantities also exceed the supply of organic matter required for denitrification.

Lance[21] reports that the nitrification-denitrification mechanism at Flushing Meadows was responsible for a net nitrogen removal of 30%. This happened because the nitrogen in the ammonium form was sorbed during the application period by the cation exchange complex and subsequently nitrified during the drying period. During the beginning of the next application period, however, the nitrate was leached through the soil in a concentrated peak, and the concentration of organic matter in the wastewater was too low for complete denitrification. By increasing the organic carbon concentration from 15 to 150 mg/l, Lance was able to obtain 90% nitrogen removal.

Another method developed by Lance to increase nitrogen removal was to collect the high nitrate water, mix it with two parts of secondary effluent and recycle it through the system. By recycling 20 to 25% of the recovered water, he was able to obtain nitrogen removals of 75 to 80%. Without these special management techniques, however, nitrogen removal may be limited to 50% or less of the nitrogen applied.

Phosphorus

Fixation and chemical precipitation are the primary mechanisms for phosphorus removal in infiltration-percolation systems. Although the capacity of the soil for phosphorus removal is difficult to predict, it is finite, and, for soils low in clay and organic matter, it may limit the long-term use of the site. The research work by EPA has resulted in a predictive model that can be used on 20 different mineral soils.[22] The model has been verified for at least one soil: Bouwer has measured 95% phosphorus removal after 200 feet (61 m) of travel through sandy loam, sand and gravel.[3]

Basin Characteristics

The surface of an infiltration-percolation basin should be designed to disperse the clogging material applied.[23] The two principal methods of accomplishing this dispersion are: (1) growing vegetation on infiltrative surfaces, thus providing root channels with attendant soil expansion, and (2) covering surfaces with graded sand, gravel, or mulch, thus encouraging clogging action in a matrix having larger pore spaces than the soil.

Although both methods have been used successfully, there is no general agreement on the best type of basin surface. What has proved successful at one location has worked poorly at others. Selection of the type of basin surface can be based on comparative operational studies at the infiltration site. Consideration should be given to renovative capacity and maintenance required as well as infiltration capacity. The relative merits of the two types of basin surface are presented in the following discussion.

The effect of vegetated surfaces is illustrated at Flushing Meadows. In 1969 the six basins were compared. Four were vegetated, one was left bare, and one was covered with gravel. The infiltration rate for the vegetated basins was 25% greater than that for the bare basin, and the bare basin showed a greater infiltration rate than the gravel-covered one.[3]

The advantages of vegetation are: (1) protection of soil from the impact of water droplets in areas of high rainfall, (2) additional nutrient removal if the vegetation is harvested (usually less than 5%), and (3) possible promotion of denitrification.[21] Further research is required to substantiate the thesis that increased soil carbon around roots promotes increased denitrification.

The disadvantages of vegetation include increased maintenance and lower recharge depths. Loading cycles must be adjusted to promote plant growth in the early growing season and to permit harvesting. At Flushing Meadows the water depth, which was found to be directly proportional to infiltration rates, was restricted to avoid complete submergence. Giant bermuda grass proved to be the most suitable vegetation although rice and sudan grass were also tried.[3]

The effect of nonvegetated surfaces is illustrated at the Whittier Narrows test basin, where 6 inches (15.2 cm) of pea gravel was spread over the bare soil to eliminate plant growth and decrease maintenance.[4] It was concluded that this layer was a major contributing factor to higher infiltration rates. On the other hand, studies at the Hemet and Flushing Meadows recharge sites indicated that a gravel cover was not particularly effective in maintaining infiltration capacity.[2,3] Researchers at Flushing Meadows hypothesized that the negative effect of the gravel layer on infiltration was due to a mulching action of the gravel layer and a resulting poor drying of the underlying soil.

When using a gravel cover, it has been suggested that abrupt changes in particle size should be avoided between the surface cover material and the soil at the infiltrative surface.[23] This precaution will prevent loss of infiltrative surface area as a result of blinding by large particles and will promote dispersion of clogging solids.

Mulching the basin surface with cotton gin mill waste produced a large and long-lasting increase in infiltration rates in experimental studies in California.[24] In these 1954 studies intermittent periods of spreading and drying, combined with surface raking or cultivation, proved conducive to the highest infiltration rates.

Distribution System

The distribution system consists of facilities to transfer wastewater to the site, hydraulic controls, and possible outlet facilities. For spreading basins the transmission can be by buried pipeline with alfalfa valves for inlets or open flumes. For sprinkler systems, buried or surface distribution laterals (solid set) are generally used with relatively high pressures and large nozzles.

At Flushing Meadows the transmission line was buried with concrete riser pipes and inlet boxes at the head end of each rectangular basin. Each inlet could be isolated by closing the alfalfa valve on the riser pipe. The wastewater passed through a triangular, critical depth flume for flow measurement as it entered the 3-ft (0.9 m) deep basins.[3] For vegetated basins, the depth of inundation was 7 in. (17.8 cm) while for bare basins, the depth of inundation was 13 in. (33 cm). The outlet control was another critical depth flume, which measured flow and kept the depth constant.

In the new (1975) project at Phoenix, the 15 mgd (657 l/sec) of secondary effluent will be spread in four basins of 10 acres (4 ha) each.[25] The depth of inundation will be allowed to fluctuate so that both basins (only two operate at any one time) will receive approximately 7.5 mgd (328 l/sec).

At Yuba City, California, six basins were constructed for a total of 140 acres (57 ha). The distribution system consists of buried concrete pipe with basin inlet structures located in the center of each basin. The basins are about 5 ft (1.5 m) deep with 2 ft (0.6 m) of freeboard. There are no outlet structures.

Control of Underground Flow

Infiltration-percolation systems discharge large volumes of water to underground flow. This flow may be to a permanent groundwater body or it may drain naturally into a surface watercourse. If the spread of renovated water into the groundwater basin is not desirable or if the transmissibility of the soil will not allow the desired long-term hydraulic loading, artificial control of the underground flow can be imposed. This can be done using horizontal drains if the aquifer is shallow or by using wells if the aquifer is deep.

Groundwater Recharge

Recharge systems, such as those in the Los Angeles area, rely on the soil profile to treat the wastewater and transmit it to the groundwater. The groundwater may then be used through pumping at individual wells or at centralized large-capacity wells. The spread of reclaimed water should be investigated and mapped thoroughly so that the owners of any nearby wells are aware of the extent and direction of recharged flows.

Natural Drainage into Surface Water

Infiltration systems can be located so that the renovated water moves vertically and laterally into a natural surface watercourse. The system at Yuba City for example, is located next to the Feather River, and the infiltration basins drain naturally and recharge the river. For such systems the distance and time of flow is important to assure proper renovation within the soil. In the design, care should be taken that the aquifer can transmit the renovated water without an undue rise in the water elevation below the basins. This can be checked using the approximate equation[9]

$$WI = \frac{KDH}{L} \tag{1}$$

where

W = width of infiltration area, ft or m

I = hydraulic loading rate, ft/day or m/day

K = hydraulic conductivity of aquifer, ft/day or m/day

D = average thickness of the aquifer taken perpendicular to the flow direction, ft or m

H = elevation difference between watercourse and maximum allowable water table level below the infiltration area, ft or m

L = distance of lateral flow, ft or m

Bouwer[9] illustrates the use of this and other equations with some comparative examples.

Interception by Underdrains

Perforated pipe made of polyvinylchloride or asbestos-cement can be installed to collect percolating water. Underdrains can be used where water table levels are high, where vertical drainage is impaired, or where the aquifer is too shallow for recovery wells. When used to intercept percolating water in well-drained systems with deep soil profiles, underdrains will not recover the renovated water effectively.

Underdrains may be installed at depths of 4 to 8 ft (1.2 to 2.4 m). In a grid layout the lateral perforated lines may be spaced from 50 ft (15 m) to over 200 ft (60 m) apart, depending on the hydraulic loading, need for recovery, and soil properties. Perforated drains may be controlled by valves or allowed to discharge continuously. Other types of drains include open interceptor ditches, tile lines, mole drains, and French drains.[26,27]

Interception by Wells

Renovated water can be completely intercepted by wells if the aquifer is thick enough and the wells are located properly. The drawdown of the wells can be calculated when the aquifer properties are known.[28] The 15 mgd (657 l/sec) system at Phoenix will employ recovery wells for interception of the renovated water.

MANAGEMENT AND OPERATION

Important aspects of infiltration-percolation system management include (1) hydraulic loading cycles, (2) basin surface management, (3) operational problems, and (4) renovated water quality.

Hydraulic Loading Cycles

A schedule of intermittent application of wastewater is required to maintain reasonable capacities. At Hyperion (Los Angeles, California) and Flushing Meadows, continuous applications have been attempted, but they showed a constant drop in infiltration rates and a reduction of renovative capacity.

Optimum loading cycles will depend on the primary objective of the system. For instance, an application cycle that maximizes infiltration rates may minimize nitrogen removal by denitrification. It is therefore impossible to predict the optimum loading cycle for any one system. The variation in reported loading cycles for existing systems, as shown in Table 10.4, illustrates this point. Experimentation is required to determine the best loading cycles consistent with the objectives of the system.

The resting period, which may vary from 1 to 20 days, is essential to allow atmospheric oxygen to penetrate the soil and reestablish aerobic conditions. As the surface dries, aerobic bacteria become active in organic matter decomposition and nitrification. This activity helps to break up the clogging layer and free ammonium adsorption sites on clay and humus materials. In addition, prevention of severe anaerobic conditions through intermittent loading will avoid the possibility of leaching degradable organics into the groundwater.

Application frequency may affect the renovation that results from adsorption by changing the detention time of the wastewater constituents in the soil matrix. The longer the detention time, the higher the probability of contact between wastewater constituent and sorption site and the greater the overall renovation. Resting periods may also allow sorbed ions to diffuse into interstices of the sorbent particles, which frees the external sorption

Table 10.4. Existing Hydraulic Loading Cycles[1]

Location	Loading Objective	Application Period	Resting Period
Flushing Meadows, Arizona			
Maximum infiltration	Increase ammonium adsorption capacity	2 days	5 days
Summer	Maximize nitrogen removal	2 wk	10 days
Winter	Maximize nitrogen removal	2 wk	20 days
Hyperion, California	Maximize infiltration rates	Continuous[a] 0.5 hr	– 23.5 hr
Lake George, New York			
Summer	Maximize infiltration rates	9 hr	4-5 days
Winter	Maximize infiltration rates	9 hr	5-10 days
Tel Aviv, Israel	Maximize renovation	5-6 days	10-12 days
Vineland, New Jersey	Maximize infiltration rates	1-2 days	7-10 days
Westby, Wisconsin	Maximize infiltration rates	2 wk	2 wk
Whittier Narrows, California	Maximize infiltration rates	9 hr	15 hr

[a]Abondoned after 6 months

site. Further research is required to determine, in actual field operation, the relationship between sorption effectiveness and application frequency.

Basin Surface Management

Where bare soil or gravel surfaces are used, they should be scarified or raked when solids accumulate. The frequency will depend on the quality of the wastewater. More frequent raking is needed when water with high solids content is used. Operational experience will indicate the frequency; a typical frequency is once a year. The methods used have included shaving or sweeping,[3] and disking, scarifying, and rototilling.[2]

For vegetated surfaces careful operation of the loading cycle is needed in the spring until the vegetation is well established. The surface may be harrowed on an annual basis to break up any solids buildup.

Operational Problems

Problems developed with infiltration-percolation systems are generally related to the weather, mechanical breakdowns, or wastewater characteristics. At Hemet and Westby, severe prolonged rains have caused serious problems. At Hemet the solution was temporary abandonment of the system, while at Westby the rain and an overloaded plant may lead to permanent abandonment. At Lake George the snow and ice do not present serious problems. As the basins are flooded with effluent, the ice is merely floated up 7 to 8 inches and serves to insulate the soil surface from further lowering of the temperature.

Mechanical breakdowns can be minimized by preventive maintenance; however, standby pumps and excess spreading basins are advisable. At nearly all operating sites, problems relating to wastewater characteristics, primarily clogging, have occurred when abnormally high solids content water is applied to the land. At Whittier Narrows, the basin was taken out of service until the plant effluent quality was stabilized. At Flushing Meadows, the upper portions of the spreading basins were converted to settling basins to reduce the suspended solids content.

Renovated Water Quality

The removal efficiencies of several selected systems are shown in Table 10.5. Because of the differences in secondary effluent qualities, soil types, depths of sampling, and parameters measured, it is difficult to compare these results.

Table 10.5. Removal Efficiency at Selected Infiltration-Percolation Sites[1]

Location	Depth Sampled, ft	Removal Efficiency, %				
		BOD	SS	N	P	E. coli
Flushing Meadows, Arizona	30	98	100	30-80	60-70	100
Hyperion, California	7	90	–	–	–	98
Lake George, New York	10	96	100	43-51	8-61	99.3
Santee, California	200[a]	88	–	50	93	99
Westby, Wisconsin	3	88	–	70	93	–
Whittier Narrows, California	8	–	–	0[b]	96	0[c]

[a]Lateral flow
[b]Short, frequent loading promots nitrification but not denitrification.
[c]Coliforms regrow in the soil.
Note: 1 ft = 0.305 m

Nitrogen

The importance of denitrification for nitrogen removal has been discussed. The hydraulic loading cycle can be manipulated to change the amount of nitrogen removal as studied at Flushing Meadows.[9] If the nitrogen concentration in the renovated water is critical to the design, considerable experimentation with hydraulic loading rates and cycles may be warranted.

Phosphorus

As described previously, phosphorus can be removed by a combination of adsorption, chemical precipitation, and ion exchange. At Whittier Narrows, the total phosphorus concentration after 8 ft (2.4 m) of percolation had been reduced to 0.1 mg/l as P.[4]

Trace Elements

Heavy metals and trace elements such as boron and fluoride may be retained in the soil profile or remain in the renovated water depending on the soil and the element.[19] At Flushing Meadows, copper, zinc, and fluoride were retained in the soil while cadmium, lead, and zinc concentrations in renovated water remained the same as in the secondary effluent. About one-half of the mercury remained in the soil.[9] Soil containing more iron and aluminum oxides will retain larger quantities of trace elements including boron.

The metal content of secondary effluent is usually quite low and below maximum limits suggested for drinking water because most of the metals in wastewater are concentrated in the sludge. Excessive concentrations of minor elements in effluent may result from industrial discharges or from the use of certain household products (boron) and are best controlled at the source.

Organic Matter

Although the BOD in the renovated water is essentially zero, some organic carbon remains in the water. At Flushing Meadows the refractory organic carbon in renovated water was 4 mg/l.[9] At Hemet the COD was reduced from 50 to 16 mg/l and surfactants were reduced from 2.6 to 0.35 mg/l after 8 ft (2.4 m) of travel.[2] The nature and toxicity of this remaining organic matter need to be identified.

Total Dissolved Solids

The total dissolved solids (TDS) concentration in the renovated water may remain the same or may increase over that in the applied effluent.

Two mechanisms that can cause TDS to increase are mineral weathering and dissolution of soil lime. The rate of weathering depends on the temperature, water quality, and the quantity of water passing through the soil profile. Dissolution of soil lime (calcium carbonate) depends on the pH, hardness, and partial pressure of carbon dioxide.

When the quantity of percolate is small, less salt (TDS) will be leached out of the profile than is applied, according to Oster and Rhoades.[29] On the basis of their model, this is a result of chemical precipitation in the soil. When the quantity of percolate is large, weathering and dissolution of soil lime increase, and the TDS in the renovated water may increase over the concentration in the effluent applied.

At Whittier Narrows, Reid[30] reports a TDS increase of 11% and a hardness increase of 30% at the 8-ft (2.4 m) depth. At Flushing Meadows, Bouwer[3] reports a TDS increase of 4%, which he related to evaporation (3%) and pH drop with resulting dissolution of soil lime (1%). Because evaporation has such a small effect on TDS for high-rate systems, the effect of a pH drop, caused by nitrification and carbon dioxide generation during BOD oxidation, is generally more significant.

Bacteria and Viruses

Evidence of the removal of coliform bacteria by soil is available from many sources.[1-6,12,13,24] The removal of fecal coliform bacteria in several infiltration-percolation systems is given in Table 10.6.

Table 10.6. Fecal Coliform Removal in Selected Infiltration-Percolation Systems

		Fecal Coliform, MPN/100 ml		
Location and Source	**Soil Type**	**Effluent Applied**	**Renovated Water**	**Distance of Travel, ft**
Flushing Meadows, Arizona[3]	sand	1,000,000	> 10	100
Hemet, California[2]	sand	60,000	11	8
Santee, California[5]	gravel	130,000	> 8	200

Note: 1 ft = 0.305 m

A number of laboratory and field studies on virus removal from water and wastewater have been reported. In a pilot scale study, Robeck[31] reports that 2 ft (0.6 m) of packed sand removed poliovirus from relatively clean water flowing at rates up to 4 ft/day (1.2 m/day), but that virus breakthrough occurred at higher flowrates. In a laboratory study, Hori *et al.*[32] found that at least 97% of the poliovirus applied to some Oahu soils was removed at soil thicknesses of 1.5, 2.5, and 6 in. (3.8, 6.4, and 15 cm). Other Oahu soils were less effective in virus removal. At Santee, California,[5] the infiltration-percolation system was challenged with a dose of 36 plaque-forming units of poliovirus, and none were detected after 200 ft (60 m) of travel.

More recently, virus studies have been conducted at St. Petersburg, Florida,[33] in which some 12 million plaque-forming units were applied per day to a sandy test site. Viruses were isolated infrequently at both 10 ft (3 m) and 20 ft (6 m) depths. A major rainfall event produced peaks of virus concentrations at these two depths. Virus studies are currently underway at a number of infiltration-percolation sites including Flushing Meadows.

COSTS

There are few cost data available on infiltration-percolation systems. Pound and Crites[1] report on two municipal systems, and Crites *et al.*[8] report on three industrial systems giving construction and operating costs. A summary of available system costs including two independent cost estimates for a typical 1-mgd (43.8 l/sec) system, are given in Table 10.7.

Table 10.7. Summary of Construction and Operating Costs for Selected Infiltration-Percolation Systems

Location and Source	Average Flow, mgd	Construction Cost: Year or Cost Index	Construction Cost: Cost, \$/gpd	Construction Cost: Cost, \$/ac	Operating, Cost, ¢/1000 gal
Westby, Wisconsin[1]	0.12	1959	0.08	2,500	–
Phoenix, Arizona[1]	0.67	1967	0.01	1,800	2.4
Bridgeton, New Jersey[8]	3.0[a]	1961	0.08	5,460	3.6
Salem, New Jersey[8]	1.3[a]	1972	0.03	1,330	–
Yakima, Washington[8]	4.0[a]	1964	0.03	1,500	–
Pound, *et al.*[34]	1.0	177.5[b]	0.24	16,900[c]	5.4
Tchobanoglous[35]	1.0	1900[d]	0.20	–	12.9
Yuba City, California	4.3	1972	0.08	2,330	–

[a]Industrial system

[b]EPA Sewage Treatment Plant Construction Cost Index for February 1973 (national average)

[c]Based on 14 acres (5.7 ha)

[d]ENR Construction Cost Index

Note: 1 mgd=43.8 l/sec; 1 gal=3.785 liters

As can be seen from Table 10.7, the costs of infiltration-percolation systems are extremely variable. The construction costs shown do not include the cost of land. The three industrial systems utilize sprinkler application and the municipal systems utilize spreading basins.

The estimated cost for a 1-mgd system[34] includes site clearing, low-lift pumping, distribution to basins at an application rate of 18 in./wk (46 cm/wk), basin construction, recovery wells 50 ft (15 m) deep, monitoring wells, administrative and laboratory facilities, service roads, and fencing. Individual cost curves for these components are contained in *Costs of Wastewater Treatment by Land Application.*[34]

Tchobanoglous[35] reports a similar construction cost range of $180,000 to $220,000 for a 1-mgd (43.8 l/sec) system. He proposes that a scale factor of 0.58 be used for other flowrates, according to the equation

$$\text{Construction Cost} = \$200{,}000 \left(\frac{\text{Flowrate, mgd}}{1 \text{ mgd}} \right)^{0.58}$$

REFERENCES

1. Pound, C. E. and R. W. Crites. *Wastewater Treatment and Reuse by Land Application,* Vol. I and II, Office of Research and Development, Environmental Protection Agency (August 1973).
2. Boen, D. F., *et al. Study of Reutilization of Wastewater Recycled through Groundwater,* Vol. I, Eastern Municipal Water District, Office of Research and Monitoring, Environmental Protection Agency (July 1971).
3. Bouwer, H., R. C. Rice, and E. D. Escarcega. "Renovating Secondary Sewage by Groundwater Recharge with Infiltration Basins," U.S. Water Conservation Laboratory, Office of Research and Monitoring, Environmental Protection Agency (March 1972).
4. McMichael, F. C. and J. E. McKee. "Wastewater Reclamation at Whittier Narrows," California Water Quality Control Board, Publ No. 33 (1966).
5. Merrell, J. C., *et al.* "The Santee Recreation Project, Santee, California," Final Report, FWPCA, U.S. Department of the Interior, Cincinnati (1967).
6. Aulenbach, D. B., *et al.* "Effectiveness of a Deep Natural Sand Filter for Finishing of a Secondary Treatment Plant Effluent," presented at the New York Water Pollution Control Association Meeting, January 1970.
7. Bendixen, T. W., *et al.* "Ridge and Furrow Liquid Waste Disposal in a Northern Latitude," *J. San. Eng. Div., Amer. Soc. Civil Engr.* **94** (SA1), 147 (1968).
8. Crites, R. W., C. E. Pound, and R. G. Smith. "Experience with Land Treatment of Food Processing Wastewater," *Proceedings Fifth National Symposium on Food Processing Wastes*, Monterey, California (April 1974).

9. Bouwer, H. "Infiltration-Percolation Systems," presented at the Symposium on Land Application of Wastewater, University of Delaware, Newark, November 1974 .
10. Thomas, R. E. and C. C. Harlin, Jr. "Experiences with Land Spreading of Municipal Effluents," presented at the First Annual IFAS Workshop on Land Renovation of Wastewater in Florida, Tampa, June 1972
11. Linsley, R. K., *et al.* *Hydrology for Engineers.* (New York: McGraw-Hill Book Co., 1958).
12. McGauhey, P. H. and R. B. Krone. "Soil Mantle as a Wastewater Treatment System," SERL Report No. 67-11, University of California, Berkeley (December 1967).
13. Krone, R. B. "The Movement of Disease-Producing Organisms through Soils," *Proceedings of the Symposium on Municipal Sewage Effluent for Irrigation*, Louisiana Polytechnic Institution, July 1968 .
14. Bogedain, F. O., *et al.* "Land Disposal of Wastewater in New York State," presented at the N.Y. State Water Pollution Control Association meeting, June 1973 .
15. Lawton, G. W., *et al.* "Effectiveness of Spray Irrigation as a Method for the Disposal of Dairy Plant Wastes," University of Wisconsin (1960).
16. National Canners Association, "Liquid Wastes from Canning and Freezing Fruits and Vegetables," Office of Research and Monitoring, Environmental Protection Agency (August 1971).
17. "The Cost of Clean Water, Volume III, Industrial Waste Profiles No. 6– Canned and Frozen Fruits and Vegetables," U.S. Department of the Interior, FWPCA (September 1967).
18. Thomas, R. E. and T. W. Bendixen. "Degradation of Wastewater Organics in Soil," *J. Water Poll. Control Fed.* **41**(5), Part 1, 808 (1969).
19. Reed, S. C. "Wastewater Management by Disposal on the Land," Corps of Engineers, U.S. Army, Special Report 171, Cold Regions Research and Engineering Laboratory, Hanover, New Hampshire (May 1972).
20. Environmental Protection Agency. "Evaluation of Land Application Systems," EPA-430/9-75-001 (Washington, D.C.: Office of Water Program Operations, March 1975).
21. Lance, J. C. "Fate of Nitrogen in Sewage Effluent Applied to Soil," presented at the ASCE National Environmental Engineering Convention, Kansas City, Missouri, October 1974 .
22. Enfield, C. G. and D. C. Shew. "Comparison of Two Predictive Non-Equilibrium One-Dimensional Models for Phosphorus Sorption and Movement Through Homogeneous Soils," *J. Environ. Qual.* **4**(2) 198 (1975).
23. Gotaas, H. B. "Field Investigation of Waste Water Reclamation in Relation to Ground Water Pollution," California State Water Pollution Control Board, Publ. No. 6 (1953).
24. Gotaas, H. B., *et al.* "Studies in Water Reclamation," SERL Report No. 37-13, Berkeley, University of California (July 1955).
25. Bouwer, H. Personal communication (March 1975).
26. Houston, C. E. "Drainage of Irrigated Land," California Agricultural Extension Service, Circular 504, Davis, California (December 1967).

27. Soil Conservation Service. "Drainage of Agricultural Land," Water Information Center, Inc. (1973).
28. Huisman, L. *Groundwater Recovery*. (New York: Winchester Press, 1972).
29. Oster, J. D. and J. D. Rhoades. "Calculated Drainage Water Compositions and Salt Burdens Resulting from Irrigation with River Waters in the Western United States," *J. Environ. Qual.* **4**(1), 73 (1975).
30. Reid, D. M. "Whittier Narrows Test Basin, Progress Report," Los Angeles County Flood Control District (July 1973).
31. Robeck, G. G., *et al.* "Effectiveness of Water Treatment Processes in Virus Removal," *J. Amer. Waterworks Assoc.* **54**, 1275 (1962).
32. Hori, D. H., *et al.* "Migration of Poliovirus Type 2 in Percolating Water through Selected Oahu Soils," Technical Report No. 36, University of Hawaii Water Resources Research Center (1970).
33. Wellings, F. M., *et al.* "Pathogenic Viruses May Thwart Land Disposal," *Water Wastes Eng.* **12**(3), 70 (1975).
34. Pound, C. E., R. W. Crites, and D. A. Griffes. "Costs of Wastewater Treatment by Land Application," EPA-430/9-75-003. (Washington, D.C.: Office of Water Program Operations, June 1975).
35. Tchobanoglous, G. "Wastewater Treatment for Small Communities," in *Water Pollution Control in Low Density Areas*, W. J. Jewell and R. Swan, Eds. (Hanover, N.H.: University Press of New England, 1975), pp. 389-427.

11

ENGINEERING INVESTIGATIONS FOR LAND TREATMENT AND DISPOSAL

Robert L. Sanks

Professor, Department of Civil Engineering
and Engineering Mechanics
Montana State University
Bozeman, Montana 59715

Takashi Asano

Associate Professor, Department of Civil
and Environmental Engineering
Washington State University
Pullman, Washington 99163*

A. Hayden Ferguson

Professor, Department of Plant and Soil Science
Montana State University
Bozeman, Montana 59715

INTRODUCTION

The purpose of this chapter is two-fold: (1) to provide a comprehensive overview of engineering for land application projects and (2) to fill in the gaps of information and data not covered by the previous chapters. A suggested reading list is appended. Most of the literature is heavily biased in favor of land application, and many readers may inadvertently develop polarized attitudes. Engineers, who are in a position of trust, owe it to themselves and to their clients to read contrary opinions and to evaluate all factors before deciding for or against land application.

*Formerly, Department of Civil Engineering and Engineering Mechanics, Montana State University.

Figure 11.1. Exposed solid-set spray irrigation system in forest at Rocky Boy Indian Reservation. Designed by Christian, Spring, Sielbach, and Associates with

the authors as special consultants. This is one of the first two spray irrigation projects for sanitary wastes in Montana.

The application of liquid wastes to the land has been practiced in the United States for many decades and in parts of the rest of the world for many centuries. Near the beginning of this decade it received renewed impetus because of the Federal Water Pollution Control Acts. The existing federal regulations, Federal Water Pollution Control Act Amendments of 1972 (Public Law 92-500), establish two general goals: (1) to achieve wherever possible by July 1, 1983, water that is clean enough for swimming and other recreational uses, and clean enough for the protection and propagation of fish, shellfish, and wildlife, and (2) by 1985 to have no discharge of pollutants into the nation's waters. The act requires that land application should at least be investigated as an alternative to water disposal, and to this there can scarcely be any objection. Certainly it is wise to make more intensive use of available water resources. But because the water to be returned to the soil is return flow from domestic and industrial uses, it is important that (1) the ability of the soil to alter water quality be understood, (2) long-time injurious or damaging synergisms be wholly prevented, and (3) no important detail be overlooked by engineers in pursuing their investigations.

PRELIMINARY INVESTIGATIONS

Preliminary investigations for the purpose of establishing the general feasibility of a site can be conducted at minimum cost without extensive laboratory or field work, and in some circumstances by inspection only if the combined suitability of soil, water, topography, and location is evaluated by experts. In general, the problems of greatest importance are (1) the achievement of optimum quality improvement by the soil to prevent groundwater pollution, and (2) the achievement of maximum hydraulic application rates consistent with the protection of both soil and crops and the elimination of the hazard of direct overland runoff into streams.

Wastewater

There need be little concern about trace chemicals in domestic wastewaters that contain little or no industrial waste. In many areas, however, boron should be investigated because soils are very sensitive to it and it is contained in some soaps. For the most part, BOD and suspended solids will be too low for concern except, perhaps, in meeting arbitrary regulations. Some surface and many groundwaters are high in dissolved solids, and care should be taken to prevent the buildup of salts in the root zone. This means the hydraulic loading of any application must be high enough to overcome the effects of evapotranspiration by ensuring adequate leaching.

The complex interactions of soil, plants, water, and climate preclude the establishment of a single criterion to evaluate all water quality requirements for land disposal. However, the Sodium Adsorption Ratio (SAR) is probably the single most important parameter in terms of prolonged water application to soils. It is defined as

$$\text{SAR} = \frac{\text{Na}}{\sqrt{(\text{Ca} + \text{Mg})/2}}$$

in which the concentration of the ions is expressed in milliequivalents/liter. Many soils and crops can accept rather high concentrations of salts provided care is taken in irrigation and the SAR is sufficiently low. But a high SAR means excessive sodium, which, sorbed by clay, causes the clay to swell and makes the soil impermeable to air and water.

An SAR of 15 in either soil or water is usually considered the breakpoint for successful long-term application of waters to soils containing significant quantities of clay. Fifteen per cent of clay, especially swelling clay, should be considered significant. However, the permissible SAR value also depends on the type of clay. Swelling clays, such as bentonite, are troublesome with an SAR value of 8 to 10, whereas nonswelling clays can tolerate SAR up to 20. When the SAR is extremely high, virtually all soils will be affected. For example, the SAR of water from an ion-exchange softener may exceed 50, and all soils other than pure sands or coarse silts are very apt to swell enough to reduce permeability.

If SAR values either in soil or in water exceed those mentioned, gypsum or some other inexpensive soluble form of calcium can be added to reduce the swelling of soil with its consequent disastrous effect on permeability. However, if it is necessary to add soluble calcium salts, strong consideration to alternative methods of wastewater disposal should be considered on the basis of cost of chemicals and operational difficulty. Even if the SAR value is within allowable limits, water low in salt but high in bicarbonate (alkalinity) may impair the permeability of soils.

The salt content of the soil solution results from natural dissolution of soil minerals, from the salt content of the irrigation water, and to some degree from fertilizer. Evapotranspiration may significantly increase the concentration of salt as water is removed from the soil. Adequate applications of irrigation water are necessary to leach salts below the root zone. Plants vary in their tolerance to salinity and discussions may be found in *Water Quality Criteria*[1] and *Agriculture Handbook 60.*[2] In general, waters having an electrical conductivity of more than 2 millimhos/cm are detrimental to the growth of all but salt-tolerant crops. Hence, electrical conductivity should be measured for all water tested for SAR.

Boron is required in trace amounts as a plant nutrient, but it becomes toxic at levels above about 1 mg/l. Eaton and Wilcox[3] found that crops typically could tolerate 0.75 mg/l in irrigation water continuously and as much as 2 mg/l for short-term applications but only on fine textured soils.

Soils

Soils information for almost any area in any state is available from the U.S. Soil Conservation Service (with a branch located in each state) or from the state land grant university. This information is often in the form of publications, usually on a county basis. The published reports contain invaluable information including maps of soil types, texture, slope, depth to water table, depth of soil, estimate of clay type, chemical characteristics, and engineering interpretations. Even if the information is not published it is usually available, although it may be somewhat more difficult to obtain.

SAR Test for Soils

SAR tests of soils are made by mixing soil into mud with just enough distilled water to glisten on top of the mud, which is placed in a Buchner funnel. The funnel should be covered to inhibit evaporation. After standing for 24 hours, suction is applied to the Buchner funnel and a sample of filtrate is obtained for analysis.

Plants as Indicators

Plants are good indicators of soil conditions, and a plant and soil scientist can, merely by inspection, tell a great deal about the salinity of the soil, its water-holding capacity, permeability, depth to water table, alkalinity, nutrient supply, soil texture, and suitability for a wastewater application site. Conclusions based on these observations must eventually be confirmed by adequate testing, but for preliminary investigations such examinations are both valuable and inexpensive. Table 11.1 is a selected list of many (though not all) of the soil characteristics that can be determined by the appearance of the plants. The appearance of the surface condition can be interpreted as shown in Table 11.2 and an inspection of the soil horizons can be interpreted as shown in Table 11.3

Field Measurements

For preliminary investigations, a visual inspection of the site by a highly qualified soil specialist (who could be a plant and soil scientist, a geologist,

Table 11.1. Probable Soil Characteristics Indicated by Plants

Plant Species	Probably Indicates
Alpine fir	Poorly drained soil, high water table
Spruce	Poorly drained soil, high water table
Cattails	Poorly drained soil, high water table
Sedges	Poorly drained soil, high water table
Willow	Poorly drained soil, high water table
Dogwood	Poorly drained soil, high water table
Needle and thread grass	Light textured, sandy soil
Western wheat grass	Heavy textured, poorly drained
Salt grass	Highly saline soil
Mexican fireweed	Highly saline soil
Grease wood	Highly saline soil, sodium problems
Foxtail	Salt, sodium, high water table
Ponderosa pine	Dry soil
Good sage brush	Good and deep soil

Table 11.2. Probable Conclusions Based on a Soils Inspection

Characteristic	Probable Conclusions
Surface	
Round spots with no or poor plant growth	Sodium problems
White salt crusts	High salinity
Salt crusts with black or brown colored layer on surface	Sodium problems
Subsurface	
Mottled color in soil horizon	High or fluctuating water table

Table 11.3. Soil Class by Drainage, Color, and Water Table

Soil Class	Color	Water Table Conditions
Well-drained	Bright red, yellow, and brown; free of mottles to a depth of 36-42 in. or more	Water table commonly greater than 60 inches during most months
Moderately well-drained	Uniform red, yellow or brown colors in the surface and upper subsoil horizons; mottles present in the lower subsoil and parent or underlying materials	
Somewhat poorly drained (imperfectly drained)	Generally mottled directly below surface horizon; color matrix below surface layer dominantly yellow and brown with gray, rust brown, and orange mottles	Mottled conditions indicate presence of high water table during some parts of the year, usually spring, late fall and winter months; seasonally, water table ranges from 24-120 inches in depth
Poorly drained	Dark-colored surface layer high in organic matter; horizon below the surface layer predominantly gray with orange, brown, rust brown or yellow mottles	High water table at or within 30 inches of surface most of the year unless artificially drained

or an agricultural engineer) is desirable. At least one test pit or auger hole should be dug to a depth of 10 feet (3.0 m), to the water table, or to bedrock (whichever makes the shallowest hole), and the soil horizons should be logged. About 2 lb (0.9 Kg) of soil should be taken from each horizon for testing in the laboratory. If the soil appears to be uniform and the area is small (less than five acres), one test pit or test boring may be adequate. However, the number of pits required is properly a function of the area, the topography, the variability of the site, and the need to find the slope of the groundwater table (which would usually require a minimum of three holes). No definite rules can be given, but even a preliminary investigation of a large site requires several holes.

Water tables in many areas are apt to fluctuate widely, and it is wise to consult a geohydrologist before digging the test pits to confirm the number required, the depth which should be attained, and any other unusual features that should be investigated.

In a few areas, the interpretation of field investigations will be so clear that no further preliminary field tests are necessary. But if there is the slightest doubt that infiltration rates might be small or might be affected by SAR, tests are required. The tests may be made in the field but are better if made on "undisturbed" cores in the laboratory because in testing a core the flow is vertical and the hydraulic gradient is defined, whereas in a field test the flow is uncontrolled vertically and horizontally and the hydraulic gradient is therefore indeterminant. There are several other reasons for the difficulty of interpreting field tests: (1) lateral and vertical nonhomogeneity of the soil, (2) difference in quality and characteristics of water in tests and the actual wastewater, (3) difference in the duration of tests and the service life of the project, (4) difference in the antecedent soil water content, (5) impermeable zones at a depth that may not influence the test but profoundly affect a large area, and (6) permeable lenses that may profoundly affect the test but may not influence percolation over a large area, and others. Without expert interpretation, infiltration over large areas cannot be modelled even approximately from tests of small areas.

The usual type of percolation tests used by sanitarians for application to drain fields (where a pit is dug and filled with water) is of no value whatever for several reasons: (1) the all-important surface of the soil is missing, (2) the water quality in the percolation test is not the same as that of the actual operations, and (3) the time of the percolation test is too short to draw valid conclusions. Healy and Laak[4] have proved the standard percolation test cannot yield valid results, and Winneberger and Klock[5] stated test pits in close proximity often yielded results differing by several hundred per cent. On the basis that no information is better than erroneous information, the standard percolation test should be avoided.

For field tests in uniform, massive soils, the double ring method is reasonably adequate. It consists of two coaxial cylinders, one about 24 inches (61 cm) in diameter and the other about 8 inches (20 cm) in diameter. They are both driven about 4 inches (10 cm) into the earth and nearly filled with water, the water levels in each always kept approximately the same. To minimize evaporation the cylinders should be covered, and the water level should not be permitted to drop to the soil surface. Measurements of subsidence are made only on the inner cylinder. The purpose of the outer cylinder is to minimize the spreading of the hydraulic flow net from the inner cylinder. The water used must correspond closely to the SAR, the total dissolved solids content, and the alkalinity of the wastewater. It is best to use the wastewater itself if practical. In order to prevent wastewater from becoming putrid, it can be dosed with 50 mg/l of chlorine and 2 or 3 mg/l of mercuric chloride. The chlorine will quickly kill most of the microorganisms and the mercury will prevent reestablishment of bacterial life when the chlorine has dissipated.

Another valid approach is to use a large (at least 6 ft x 6 ft) metal dike to make a deep lysimeter. The edge effects are minimized by the large size, especially in a long term run, and adequate data may be obtained in reasonably uniform soils.

With most soils short-term infiltration runs are of little value in determining long-term acceptability of high rates of water. Even with small samples in the laboratory it often requires a week or longer to reach equilibrium permeability rates; thus field tests should generally be of at least ten days duration because the reactions in a soil that may cause it to attain an unacceptably low permeability are time-dependent. These reactions might be: (1) a general tendency to swell, (2) a reduction of soil salt content because of the leaching, (3) a high SAR of the wastewater, or (4) the time required for the water to reach an impermeable barrier or hydraulic (and organic) loading from the sewage. One might be able to determine if the latter is the cause by allowing the soil to drain for a period of time (two weeks, for example) and run the test a second time. If the short-term (a couple of days) flow rate is nearly the same as the similar time period flow rate of the first test, swelling is not the problem and the test results can be used with caution to establish a preliminary value of the hydraulic loading capacity of the soil. If the swelling is responsible for the reduction in infiltration rate, calcium added to the sewage can determine the adequacy of this means of control. As an alternative, the tests can be continued long enough to determine the long-term hydraulic loading capacity of the soil after it has established an SAR salt equilibrium with the water applied. Infiltration rates are usually greatly increased by alternate wetting and drying periods although the amount of increase can only be determined by testing.

Field infiltration tests are generally expensive. Inspections, measurements, and additions of water must be made at frequent intervals and the amounts of water required are considerable. If salts are to be added in a carefully controlled manner, the water can become expensive, and it is expensive to keep a crew in the field or to send them into the field at frequent intervals. Vandalism also might be a factor. Moreover, field test results show great dispersion even in apparently uniform soils. Single tests can never be trusted. Consequently, laboratory measurements on undisturbed test cores are often preferred because: (1) they are less expensive, (2) replicates give much better correlation, and (3) the results tend to be conservative since lateral movement of water (which may negate field tests) is prevented. The disadvantage is the requirement of a device to obtain the cores.

Laboratory Measurements

The laboratory measurements should include particle size distribution for each soil horizon, swelling characteristics, and the moisture-holding capacity of the soil. The test procedure for particle size distribution is given by ASTM D422-63.[6] The percentage of clay (particles smaller than 5 microns) is the most significant part of this test. Swelling can be determined in many ways, some of which are sophisticated and expensive. However, it is satisfactory to pack a sample of the soil into a 2-inch (5.0 cm) diameter glass tube to a depth of 1 inch (2.5 cm) and allow it to imbibe water (preferably the wastewater) from the bottom. Observe the height of the soil after swelling. If the amount of swelling is more than 30% of the original height, the soil will be troublesome, and the site should be rejected.

For preliminary investigations, the moisture-holding capacity can be adequately determined by tamping the soil lightly into a 1-in. (2.5 cm) diameter glass or lucite column and allowing percolating water to consolidate the sample. The moisture-holding capacity can be calculated crudely as the weight of water in the pores of the drained column.

The best test for hydraulic conductivity is a saturated permeability test made on undisturbed soil samples. Undisturbed cores can be obtained with the apparatus shown in Figure 11.2. The soil column should be at least two feet (61 cm)in depth, and greater depth is desirable. The undisturbed sample must always include the soil horizon likely to prove least permeable. As with field measurements of infiltration, the laboratory columns should be run with the same kind of water as the wastewater or, preferably, with the wastewater itself. Again, the soil should be tested for at least 14 days to determine whether the SAR values will prove troublesome. If the hydraulic conductivity rate decreases, investigations of the effect of alternate wetting and drying periods and the effect of calcium addition should be made as previously described.

The one potentially serious problem with this method is the compaction that may occur during sampling with some soils, especially at high water content. This compaction can easily be evaluated by comparing the length of sample to the hole depth. Compacted samples should be discarded. The system cannot be used on soils containing stones that prevent sampling.

The wastewater should be tested for salinity and specific ions. This can be done by measuring ammonium, sodium, potassium, calcium, magnesium,

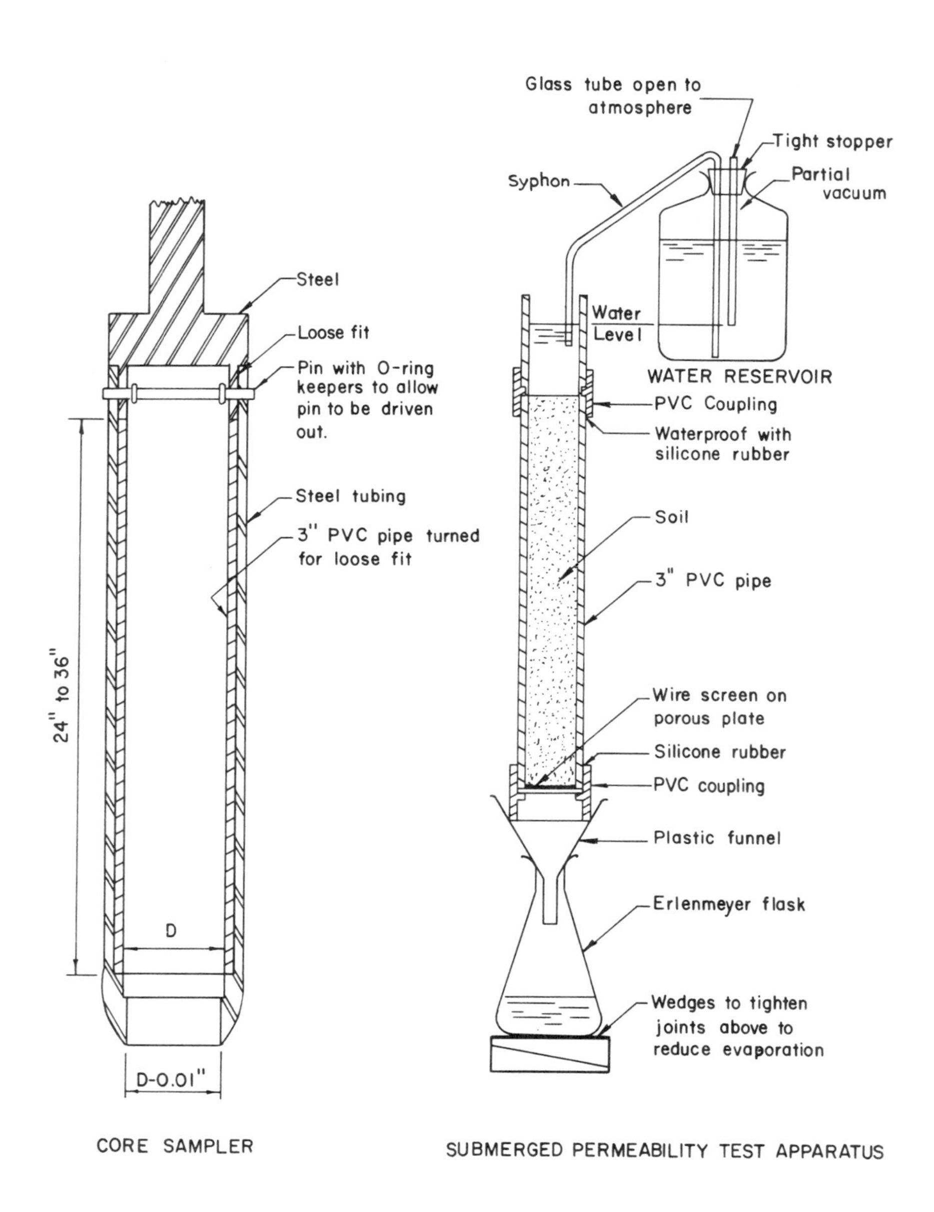

Figure 11.2. Apparatus for taking and testing undisturbed cores.

chloride, sulfate, and nitrate, which constitute all the major ions found in wastewater. An approximate means is to determine the Total Filterable Residue according to *Standard Methods*;[7] still another method is to measure the electrical conductivity of the water and compare it with tables of conductivity versus salinity found in *Agriculture Handbook No. 60*.[2]

Rejection

Any one of the following disadvantages of the site is cause for potential rejection:

1. slopes greater than 15% in forested areas and greater than 10% in grassland (under certain conditions there may be exceptions to these values depending on soil permeability and ratio of application)
2. bedrock closer than 3 ft (91 cm) to the surface or extensive outcroppings of bedrock
3. soils that swell more than 30%
4. infiltration rate results that indicate either extremely low rates or trouble with SAR values
5. water table closer than 3 ft (91 cm) to the surface during the spraying season [EPA may require a minimum of 5 ft (1.5 m) during spraying periods]
6. highly saline soils or highly saline wastewater
7. excessive boron or copper in the wastewater
8. several barely acceptable individual characteristics listed above, which in combination indicate an unsuitable site.

Pitfalls

Short-term tests discussed in the preceding section are inadequate for assessing the effects of organic and bacterial clogging, nor can short-term tests reveal the difficulties following conversion of an aerobic soil to an anaerobic one. Natural precipitation and common irrigation practice provide resting periods during which soil moisture moves downward, drawing air into the soil. The soil thereby becomes generally aerobic (although it contains many tiny anaerobic pockets). If irrigation is too heavy and frequent, the soil becomes entirely anaerobic with serious troubles to follow: odors, poor quality of percolate, and an increase of anaerobic bacteria, some of which have gelatinous sheaths that greatly reduce permeability. Hence, resting periods must be scheduled at frequent intervals.

The organic load in domestic sewage when applied at reasonable rates is very small compared to the organic content of soil, so if aerobic conditions are maintained the added organics are of little consequence and the hydraulic conductivity of the soil is not likely to be greatly reduced. But engineers are driven by economics to apply the maximum amount of wastewater possible, and all too frequently design infiltration rates have been set with more enthusiasm than sense. A high percentage of projects have been failures because long-term infiltration rates could not match predictions from short-term tests. Pierce[8] described a small Michigan project in which soil borings indicated the average application rate could be 2 inches (5.0 cm) per week and in dry weather 4 inches (10.2 cm) per week. Water was stored during the cold season and applied during a 7-month period at an average rate of 2 inches (5.0 cm) per week. Before the end of the first year, it was clear that the soils could not accept such a high application rate, not even (Pierce speculated) if good underdrains were to be installed.

Reporting on another test, Rhindress[9] stated that several years of application of an industrial waste were required to clog a certain spray irrigation field completely and to ruin the field. He reported that just applying an added increment of water to a soil is a stress and requiring the soil to act as a treatment device is another stress. When too much stress is applied over a long time, the ecosystem is gradually altered and may eventually be destroyed. These effects take place over several years and are not apparent in hydraulic conductivity tests.

The message, clearly, is that engineers should be cautious in establishing design application rates. Of course, the design application rate is not based directly on hydraulic conductivity tests anyway; it is based instead on a complete analysis of geohydrological conditions of which the hydraulic conductivity is only a part. Refer to Warner (Chapter 6) for a detailed discussion. The two worst pitfalls according to Pierce[8] are excessively optimistic infiltration rates and failure to consider the necessity for subdrains. These errors in judgment lead to ponding, nuisance, high water tables, the need to purchase more land and extend the irrigation system, and the need to install drains at considerable and unexpected expense.

It is well to consider the difference between wastewater disposal and irrigation. The objective of wastewater treatment and disposal is to apply as much water as possible on as small an area as possible to reduce costs. On the other hand, the objective of irrigation is to use as little water as possible (to reduce costs) and cover as much land as possible (to increase profits). The difference itself in viewpoints is a pitfall if not recognized and if precautions are not taken. Other pitfalls are given by Lee (Chapter 9).

PRELIMINARY DESIGN CONSIDERATIONS

Climate

While some spraying can be done under subfreezing conditions and on shallow snow cover, the sublimation of ice and snow is low, as must be the application rates. Furthermore, the danger of a thaw and excessive surface runoff from heavy snow or ice cover into water courses is ever present. Hence, prudence dictates the use of storage lagoons to contain the wastewater until the conditions are proper for spraying. Northern regions are usually frost-free for at least five months, although during some of this time, the ground may be saturated with rain and in no condition to accept additional water. Therefore, storage periods of not less than 180 days must usually be provided in severe climates such as Montana or North Dakota. However, allowance must be made for late thaws and early frosts and also for combinations of late thaws immediately followed by heavy rains. Data on local weather conditions should be obtained for at least the last 10 years and preferably for the last 20 or 30 years. Critical periods can be selected by examination and these should be studied in depth. In subtropical climates, some storage is needed during wet seasons.

Storage Lagoons

The design of a storage lagoon can be based upon a mass diagram similar to that of reservoir design. The mass diagram should take into account the inflow of the wastewater together with its monthly fluctuations, evaporation losses from the water surface, the evaporation of the spray itself (as much as 15%), and evapotranspiration of the crops. Plots of the mass curve will indicate the total required volume of storage. The required withdrawal can be determined from the number of times the wastewater can be sprayed and the total volume of water to be discharged during the season.

Location

Spray fields and lagoons should be located in the prevailing downwind direction from towns or habitation and at least far enough away so that problems with odors and aerosol pathogens do not arise. Data on wind can be obtained from the weather bureau and from interviews with local inhabitants. As wastewater disposal systems will normally include storage lagoons, spraying can be avoided on days when winds are unfavorable, or irrigation can be scheduled for windward sections of the spray field. Aerating the storage lagoons minimizes odor and chlorination reduces pathogens, but total absence of odor and microorganisms cannot be assured.

Even in preliminary investigations, the engineer should consider slope, suitability of soil, vegetation, movement of groundwater and the change of direction of groundwater movement (due to the water table mound created by spraying), wells (which might be affected by alterations of quality of groundwater and the implications of liability and lawsuits), adequate buffer zones for drifting spray, and available adjacent land area for extending the spray field if required.

Salinity

Evaporation, evapotranspiration, and sublimation have important effects upon land application. Evaporation from lagoons considerably reduces the amount of water that must be sprayed, but at the same time the salt content of the wastewater, already raised by 150 to 300 mg/l by community use, increases. In some areas the salt content may already be critical. Evapotranspiration, evaporation, and sublimation thus further increase the salt content. Unless the amount of water that is applied in a spray period is sufficient to wash these salts downward into lower soil horizons below the root zone or into the groundwater table, the salinity of the soil will become a serious problem. Evaporation data are usually available from the weather bureau. Transpiration can be computed by a number of methods of which the method of Blaney, *et al.*[10] is probably the easiest.

Hydraulic Loading

The literature contains numerous references to extremely high rates of hydraulic loading. Such rates can be applied when the soil is sandy and both the infiltration and percolation rates are high and unaffected by substances in the wastewater. Unless conditions are extremely favorable for high rates and unless nitrogen removal is of no consequence, consideration should be given to low rates of loading in the range of 0.5 to 2 inches per week. Some investigators have stated these low rates are the only ones that are safe for long-term use. The escape of nitrogen into the groundwater may be the limiting factor for hydraulic loading. If BOD-5 is less than 1000 mg/l, hydraulic loading rather than BOD loading will govern application rates because soil (which is rich in living organisms) can accept and oxidize high application rates of organic material.

Typical wastewater land application rates, together with other characteristics of spray irrigation, are summarized in Table 11.4. It should be understood, however, that these application rates should be modified by the conditions of plants, soils, and climate determined by the site investigations.

To maintain aerobic conditions in the soil, each spray application should be followed by a minimum of two or three days rest. During this time

Table 11.4. Summary of Hydraulic Application Rates and Other Characteristics of Spray Irrigation Systems[a]

Characteristic	Range
Liquid loading rate[b]	0.5-4 in./wk
Annual application	2-8 ft/yr
Land required for 1-mgd flow	140-560 acres plus buffer zones
Application techniques	Spray or surface
Soils	Moderately permeable soils with good productivity when irrigated
Probability of influencing groundwater quality	Moderate
Needed depth to groundwater	About 5 ft
Wastewater lost to:	Predominantly evaporation or deep percolation

[a]After Pound and Crites[11]

[b]Irrigation rates of 4 in./wk are usually seasonal; yearly maximum loads of 8 ft/yr would average about 2 in./wk

water percolates downward through the soil, which allows air to be drawn into it. The next application of water tends to force the column of air downward, thus providing the aerobic conditions required. Furthermore, a drying period is necessary to prevent the buildup of bacterial slimes that clog the surface and reduce infiltration rates.

Soil, especially soil containing significant amounts of clay, is a highly effective combination filter-biological digester. Virtually all suspended solids and most organic materials added to most soils, especially soils with vigorous plant growth, are likely to be removed from the percolating water.

Almost any soil has a tremendous ability to host the microorganisms for organic carbon removal provided aeration and nutritional requirements of the microorganisms are met. With an active microbiological community, it is extremely unlikely that significant amounts of organic carbon will increase except perhaps as rather stable humus. Except in extremely porous soils, little organic carbon or BOD tends to move through the soils with the water.

Materials such as heavy metals and phosphates do not move readily through the soil and are strongly sorbed on the clay and on organic colloids.

Multivalent cations tend to be retarded by the cation exchange complex of the soil, especially under slightly alkaline conditions. Anions that are not chemically sorbed (such as nitrate) move readily with the water through most soils. Nitrate is often found in high concentrations in highly treated wastewater, and it is a most serious pollutant of groundwater. Fortunately, nitrate is readily taken up by plants providing the rates of application are not too high, and, further, there may be some loss of nitrate by denitrification. But even under the highest application rates, a significant portion of nitrate is removed from the percolating waters. Estimates of nitrogen and phosphorus uptake by various crops are given in Table 11.5.

Table 11.5. Estimates of N and P_2O_5 Uptake by Crops with Good Yields[a]

Crops	N, lb/ac	P_2O_5, lb/ac
Corn	160	60
Wheat	70	30
Sugar beets	150	60
Potatoes	225	80
Alfalfa and grass	170	45
Clover	80	20

[a]After Romaine[12]

FINAL INVESTIGATIONS AND DESIGN CRITERIA

After a suitable site has been selected upon the basis of the preliminary investigations, it is necessary to obtain more accurate data on infiltration and percolation, on the fate of nitrogen and other pollutants, and on the accumulation and movement of water. A highly qualified soil scientist and a geohydrologist should be consulted.

If preliminary infiltration rate tests were made in the field, they should be verified by saturated permeability tests on an adequate number of undisturbed cores. The number and location of undisturbed cores and test holes to be drilled depends upon the size of the site, the topography, geology, and other factors. Because 10-fold variations of permeability in a small area are common, at least five drill holes and five permeability tests should be used to characterize a small site. Considerably more should be used for large sites. Because of high charges for move-on and move-off, several added drill holes and test pits do not greatly increase the cost of the investigations.

A reasonable number of drill holes for permeability tests can be determined by

$$\text{number of holes} = 4 + (\text{area in acres})^{0.5}$$

but visual observations of such factors as slope and vegetation generally indicate major soil differences in an area and this affects the number of holes required.

The data obtained for each location should include: (1) depth and soil type for each horizon of soil, (2) percentage of clay, (3) swelling characteristics for clay soils, (4) depth to bedrock or hard pan, and the condition of the bedrock (shattered, massive), (5) elevation of water table and, if possible, the variation of depth of water table in different seasons of the year, (6) groundwater pH and conductivity, (7) SAR ratio of the various soils, and (8) infiltration or hydraulic conductivity test data.

Infiltration

The data obtained from all sources must be sufficient for the soil scientist, the engineer, or the geohydrologist to establish infiltration rates that can be maintained from year to year in spite of variations in climate and the changes in soil characteristics. Any low safety factor should be counterbalanced by plans for utilizing more land and extending the spray field.

Hydraulic loading must be based upon considerations of nitrogen removal, salinity control, effect of existing and future SAR, crops, harvesting, and height of groundwater table. The hydraulic applications must be low enough to leave most of the nitrates in the root zone for removal by the plants and yet high enough to leach excessive salts to horizons below the root zone. Nitrate is one of the most serious pollutants of groundwater and care should be taken to minimize groundwater contamination. The importance of proper hydraulic loading rates cannot be overemphasized.

Groundwater

Enough data must be obtained to allow the geohydrologist or the engineer to determine the disposition of the applied wastewater. High TDS and SAR of the wastewater may create a nuisance which should not be allowed, nor should the groundwater mound under the site be permitted to rise close to the surface. There should always be at least five feet of unsaturated soil available before the water enters rock fractures or porous formations.

The movement and direction of flow of groundwater is likely to be altered by spray irrigation and the geohydrologist or engineer must protect well owners from undesirable effects at the beginning of operation and also after the final flow regimen is established. Detailed investigation of groundwater movements, bedrock, and soil conditions are required.

Both soils and bedrock conditions are so variable on this continent that generalizations are valueless. Each site is likely to be greatly different from others, so each requires individual investigation.

Design Criteria

The design begins with the consideration whether to spray the wastewater effluent as it is produced or whether to provide treatment and storage, and if so, to what extent. Even if year-around spraying were feasible, prudence dictates some storage is desirable if only to permit servicing equipment and to prevent a disaster in the event of equipment failure. The problems of runoff with snow melt, low rates of sublimation of ice and snow, freezing pipelines in cold weather, and saturation of soil during and after rain all point to the desirability of spraying only during the relatively dry period of the year. In cold climates, the spray season is not likely to exceed six months, so that lagoons with a total storage capacity to hold late autumn, winter, and early spring wastewater flows are needed.

Lagoons

Unless a means can be found to take water only from the aerobic, upper layers of a facultative pond, some odor will result from spraying, and it could be severe. Furthermore, it seems likely the EPA will require 85% removal of BOD (comparable to secondary treatment) before application to land; this might be attainable only by aerating the lagoon, particularly if the surface freezes in the winter. However, in the authors' opinion the concentration of BOD encountered in domestic wastewater is of little importance because the soil usually removes at least 95% of the BOD anyway.

In porous soils, linings are required to prevent seepage. Clay blankets, plastic and rubber sheeting, and shotcrete have all been successfully used. All linings must be protected from uplift hydrostatic pressures when the lagoon is partially dewatered, and this means the lining must be above the groundwater table. Sand blankets beneath the lining may be required for areas or horizons of semipermeable soils. Rubber and plastic membranes must be covered with soil and the banks must be graveled to protect them from wave action. The bottom must be protected from scour depending upon the type of aerator used.

Aerators

There are many satisfactory devices for aerating lagoons to maintain aerobic conditions, but storage lagoons for spraying are subject to large changes in storage volume, so the aerators used must be compatible with fluctuating depths. Surface aerators, for example, are not satisfactory unless the bottom is protected from scour when the water level is low. Suitable systems include Air-Aqua,[a] Helixor,[b] Aero-Hydraulics,[c] and Kenic[d] among others.

Good treatment is not easily obtained in single lagoons. The writers recommend designing a two- or three-unit system consisting of a preaeration lagoon of short (5 to 10 days) detention time followed by one or more storage lagoons. Most of the BOD will be removed in the preaeration lagoon, which acts as a flow-through extended activated sludge plant. It ought to be designed for constant water depth, and it must have an adequate supply of air. The storage lagoon merely retains the effluent from the preaeration lagoon in the horizon between depths of five feet minimum for Helixors or two feet minimum for Air-Aqua, and whatever the designer chooses as a maximum depth. Such lagoons require only enough air to maintain a dissolved oxygen concentration in excess of 1 or 2 mg/l and to provide slow mixing.

All of the features of recommended practice should be incorporated in the lagoon design. Guidelines are given in the *Recommended Standards for Sewage Works*[13] commonly known as the "Ten State Standards."

Spray Systems

Norum (Chapter 12) has included both design considerations and cost estimates in his discussion. The information given in Chapter 12 on design may be augmented by Pair *et al.*,[14] which appears to be the only suitable text available on this subject.

Ward[15] recommended nonclogging fog nozzles be mounted as high as possible to allow contact with trees and foliage. Two nozzles with desirable characteristics are the Bete[e] and the Agrijet.[f] However, according to Norum[16] high-pressure nozzles produce fine spray that is carried for large distances, even in light winds. If wind drift is a problem, it can be overcome by the use of a center pivot system and low-pressure sprays, which

[a]Hinde Engineering Company, 654 Deerfield Road, Highland Park, IL 60035.
[b]Polcon Corporation, Suite 305, 222 Cedar Land, Teaneck, NJ 07666.
[c]Aero-Hydraulics Corporation, 300 Place d'Youville, Montreal 125, Canada.
[d]Kenic Corporation, One Southside Road, Danvers, MA 01923.
[e]Bete Fog Nozzle, Inc., 305 Wells Street, Greenfield, MA 01301.
[f]Melnor Industries, Irrigation Division, 547 N. Citrus Avenue, Covina, CA 91723.

produce large drops that fall straight to the ground (see Figures 12.10 and 12.11). Such center pivot systems are made by McDowell[a] and Lockwood,[b] among others. Norum also stated that application rates of wastewater effluent should be much less than for normal irrigation because normal irrigation is applied only when the soil is dry, whereas sewage is applied after short resting periods when the soil still contains considerable moisture.

Hillsides and Slopes

Sepp[17] discussed the problems of spraying on hillsides and concluded that slopes up to 15% offer no particular difficulties. The more conservative recommendations of the Spray Irrigation Manual[18] are: maximum acceptable slope for unsodded fields is 4%, for sodded fields is 8%, and for forested lands is 8% for year around operation or 14% for seasonal application.

Sludge

The disposal of sludge upon land has been practiced for many years. In wastewaters without toxic metallic ions, there seem to be no serious problems with disposing of sludge on the land in uninhabited areas. With proper spraying operations, this sludge will simply add to the mulch on the land surface, and it could be beneficial rather than detrimental. Dotson[19] has discussed this topic. However, EPA requirements for pretreatment may not be compatible with the spraying of sludge together with wastewater. On the other hand, since sludge is commonly applied to land, a small area of the spray field might be set aside to receive only concentrated sludge (and lower hydraulic loadings) after which the sludge might be disked (or otherwise incorporated) into the soil if it is necessary to maintain high infiltration rates or to eliminate nuisance.

Health Protection

Lagoons are excellent devices for the reduction of bacteria because of predation, competition, and starvation. Although epidemiological studies have not shown disease to be transmitted by spray from sewage, it is known that pathogens can persist for a long time when spray is carried by wind. It may be prudent to disinfect the treated effluent to the same standards required for discharge to streams. Chlorination is effective against naked bacteria, and it seems logical to expect bacteria attached to or entrapped

---

[a]McDowell Manufacturing Company, Dubois, PA 15801

[b]Lockwood Corporation, PO Box 160, Gering, Nebraska 69341

in solid particles to fall quickly to the ground within the spray field, where they will be subjected to more predation and unfavorable environment. (Spraying can, of course, be curtailed when the winds are strong.) Dechlorination would seem to be unnecessary because sunlight and aeration dissipate chlorine rapidly.

Welch and Spyridakis (Chapter 3) stated that pathogenic organisms deposited on the ground can persist for a month or more. In order to prevent such organisms from being washed into streams by means of overland runoff in hard rains, consideration should be given to building an interception ditch downhill from the spray area and leading the runoff water from the first storm after spraying back to the lagoon. Willems[20] agreed to the desirability of an interception ditch but considered the flushing action from the first rain after spraying would make it unnecessary to intercept subsequent runoffs.

Spray areas should be fenced or posted, and a buffer zone should surround the spray field. The width ought to be: (1) a function of aerosol drift (which depends on the prevailing winds, the type of nozzles used and the system pressure) and the downwind land use (such as open range or housing). Even light airs can carry aerosols long distances. The *Spray Irrigation Manual*[18] requires a buffer zone of 200 feet, whereas the Minnesota Pollution Control Agency[21] requires a quarter of a mile. An extensive summary of health hazards is presented in Chapter 5.

Overland Flow and Infiltration Ponds

There are three other ways to apply water to land: (1) overland flow, (2) infiltration-percolation, and (3) ridge and furrow irrigation. In overland flow the water flows slowly through grass on prepared, gently-sloping land. Some may infiltrate the ground, but that is not necessary and the method is adapted to impervious soils, particularly in mild climates such as in the southern United States. A typical loading rate is 4 inches/week, and contact with the vegetation and its associated biota typically removes about 90% of the nitrogen, 50% of the phosphorus, and produces an effluent with less than 10 mg/l of suspended solids and 10 mg/l BOD-5, according to Thomas, Jackson, and Penrod.[22] Research in progress is expected to add substantially to this relatively new process, and designers should be alert to the latest findings. Some knowledge about overland flow can be gleaned from *Evaluation of Land Application Systems,*[23] Pound and Crites,[11,24,25] and Powell.[26]

Water is ponded in basins or in contour ridges and furrows in the infiltration-percolation method. The soils must be very permeable and the application rates could be very high, as much as 10 feet per week in sand.

But maintaining such rates over a period of years without ecological damage may be impossible. The groundwater quality is certain to be affected. For example, high infiltration rates at Seabrook Farms[27] has led to profound changes after several years and the system cannot be continued without modifications. With care, expert surveillance, and resting periods, however, the system is in widespread use for groundwater recharge in southern California, where high infiltration rates are used. Extra care must be exercised by a competent geohydrologist to make certain the underground boundaries of the infiltration-percolation basins can discharge the high quantities of water applied. At many sites this, and not the percolation rate *per se*, will be the rate-limiting mechanism. An extensive discussion of rapid infiltration is given in Chapter 10.

Pound and Crites[11] contend neither of the above methods are as safe as spray irrigation, both requiring more careful management than would appear justified for small communities. However, in farming communities there may be considerable operator expertise available for overland flow, and when more experience has been gained it may become a favored and economical approach in suitable areas.

Ridge and furrow irrigation, contour flooding and other forms of irrigation may have advantages, particularly in buffer zones where spray cannot be applied. Farmers and agricultural engineers are generally familiar with such methods. A summary of site selection factors and physical criteria are presented in Table 11.6.

COSTS

In general, the cost of well-designed land disposal following secondary treatment is comparable to some processes of advanced waste treatment, which explains why the resurgence of interest in land disposal had to await public insistence on advanced treatment and the willingness of the public to pay for it. Nesbitt[28] stated the cost of operation, maintenance and amortization (20 years at 6%) for a primary and secondary sewage treatment plant would be increased by about 50% by the addition of land disposal by spray irrigation. A comparison of Tables 11.7 and 11.8 corroborates Nesbitt's statement. Note, however, that neither Nesbitt's statement nor Table 11.8 includes the costs of storage lagoons or long pipelines to spray fields.

Lagoons

The relationships between average discharge and the size of lagoons required for 180 days of storage (with no allowance for evaporation) are

Table 11.6. Site Selection Factors and Criteria for Spray Irrigation[a]

Factor	Criterion
Soil type	Loamy soils preferable but most soils from sands to clays acceptable
Soil drainability	Well-drained soil preferable; consult experienced agricultural advisors
Soil depth	Uniformly 5 to 6 ft or more throughout sites preferred
Depth to groundwater	Minimum of 5 ft preferred; drainage to obtain this minimum may be required
Groundwater control	May be necessary to ensure renovation if water table is less than 10 ft from surface
Groundwater movement	Velocity and direction must be determined
Slopes	Up to 15% acceptable with or without terracing
Underground formations	Should be mapped and analyzed with respect to interference with groundwater or percolating water movement
Isolation	Moderate isolation from public preferable, degree dependent on wastewater characteristics, method of application, and crop
Distance from source of wastewater	A matter of economics

[a]After Pound and Crites.[11]

given in Figures 11.3 and 11.4. The minimum depth for lagoons aerated with an Air-Aqua system is 2 feet and only the layers above 2 feet are available for storage. The minimum depth for a Polcon system, which uses Helixors (or for other "air guns"), is 5 feet. The greatest maximum depth the site will allow must always be used to minimize icing and algae problems and to conserve land. The preaeration lagoon, especially, should be designed for the largest practical constant depth to conserve heat because low temperature inhibits biological oxidation. For small systems a buried and covered cylindrical concrete basin is worthy of consideration.

Reliable capital costs and operating expenses for the common processes used in wastewater treatment can be obtained from Patterson and Banker,[30] including cost curves for lagoons of 5, 10, and 15 ft depths, surface aerators, embankment protection, chlorination systems, and engineering, and in

Table 11.7. Treatment System Performance and Cost Estimates[a]

System	Effluent–mg/l					Total Costs–$ Million, Total Annual Costs–¢/1000 gal					
						1 mgd		10 mgd		100 mgd	
	BOD	COD	SS	P	N	Cap.	¢/1000	Cap.	¢/1000	Cap.	¢/1000
1	20	50	20	10.0	20	1.0	42	5.2	19	27	10
2	20	50	20	2.0	20	1.1	48	5.3	23	27	13
3	7	20	10	0.5	3	1.5	74	7.4	33	39	20
4	4	15	2	0.2	2	1.6	82	8.0	37	42	21
5	1	8	1	0.2	18	1.8	108	9.7	49	55	27
6	1	8	1	0.2	2	2.2	123	10.2	53	58	30

[a]After Swanson.[29]

System Description

1. Single-stage activated sludge (A.S.)
2. Single-stage A.S. and phosphorus (P) removal
3. Three-stage A.S. and P removal
4. Three-stage A.S., P removal and filtration
5. Single-stage A.S., two-stage lime, filtration and activated carbon
6. Three-stage A.S., P removal, filtration and activated carbon

Notes

A. Costs are average for U.S. (May 1971)
B. Total annual costs include operation, maintenance, and amortized capital costs (6%–25 years)
C. Performance estimates are based on well-operated municipal plants with normal strength wastewater
D. Abbrev.: BOD–5-day biochemical oxygen demand, COD–chemical oxygen demand, SS–suspended solids, P–phosphorus as P, N–nitrogen as N
E. Sludge disposal by sludge drying beds for 1 mgd and vacuum filtration and incineration for 10 and 100 mgd

Table 11.8. Comparison of Capital and Operating Costs for 1-mgd Spray Irrigation, Overland Flow, and Infiltration-Percolation Systems[a,b]

Cost Item	Spray Irrigation	Overland Flow	Infiltration-Percolation
Liquid loading rate, in./wk	2.5	4.0	60.0
Land used, acres	103	64	–
Land required, acres[c]	124	77	5
Capital costs			
Land @ $500/acre	$ 62,000	$ 38,500	$ 2,500
Earthwork	10,300	64,000	10,000
Pumping station	50,000	50,000	–
Transmission	132,000	132,000	132,000
Distribution	144,000	64,000	5,000
Collection	–	6,000	30,000
Total capital costs	$398,300	$354,500	$179,500
Capital cost per purchased acre	$ 3,200	$ 4,600	$ 35,800
Amortized cost[d]	$ 37,000	$ 34,700	$ 19,500
Capital cost, ¢/1000 gal	10.1	9.5	5.3
Operating costs			
Labor	$ 10,000	$ 10,000	$ 7,500
Maintenance	19,400	12,000	3,500
Power	5,800	5,800	1,800
Total operating costs	$ 35,200	$ 27,800	$ 12,800
Operating cost, ¢/1000 gal	9.6	7.6	3.5
Total cost, ¢/1000 gal	19.7	17.1	8.8

[a]Estimated for 1973 dollars, ENRCC index 1860 and STPCC index 192.

[b]After Pound and Crites.[11]

[c]20% additional land purchased for buffer zones and additional capacity.

[d]15-year life for capital items, excluding land; interest rate 7%.

addition, land requirements. They do not, however, give costs for waterproof linings nor for subsurface aerators. Subsurface aerators are superior to surface aerators in lagoons of fluctuating depths because scour and ice problems are minimized.

Waterproof Linings

Waterproof linings may be required to prevent unsprayed effluent with its nitrate, residual BOD and other pollutants from seeping directly into the

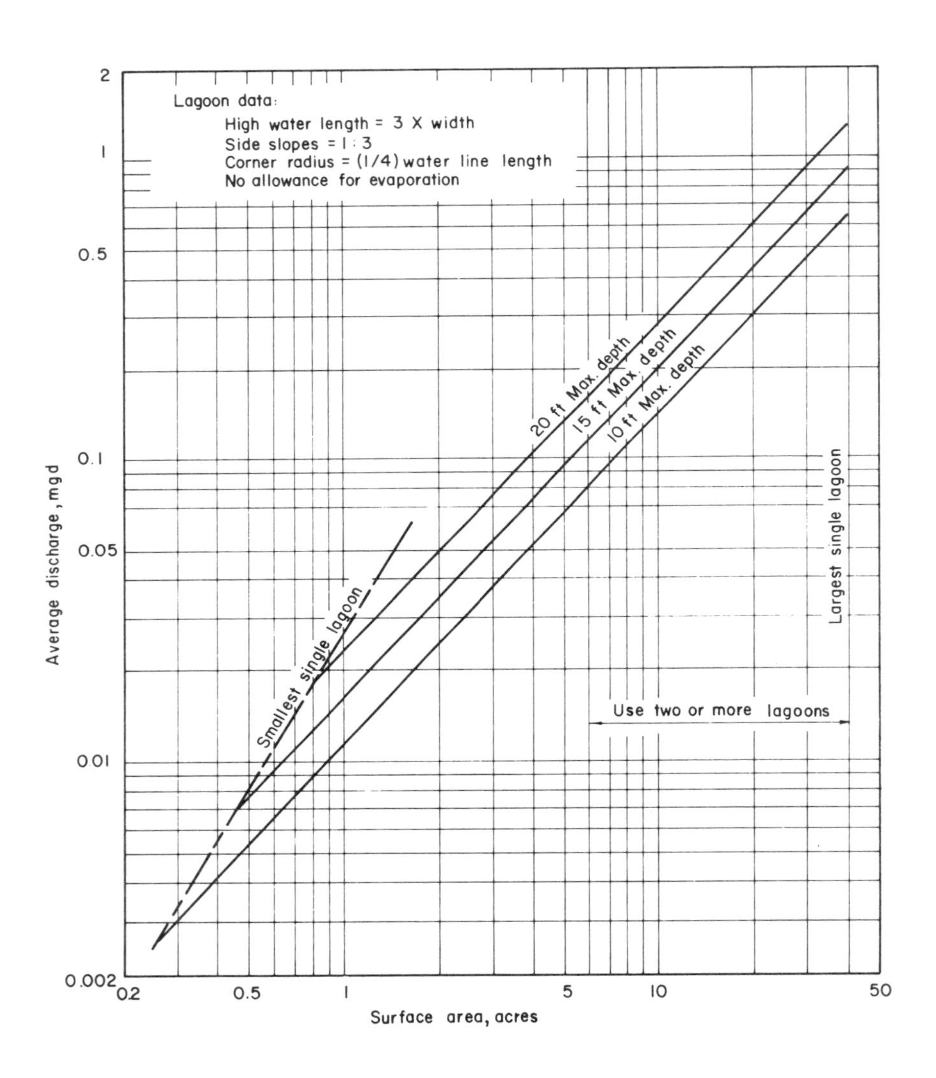

Figure 11.3. Size of single lagoon for 180-day storage and 2-ft minimum depth.

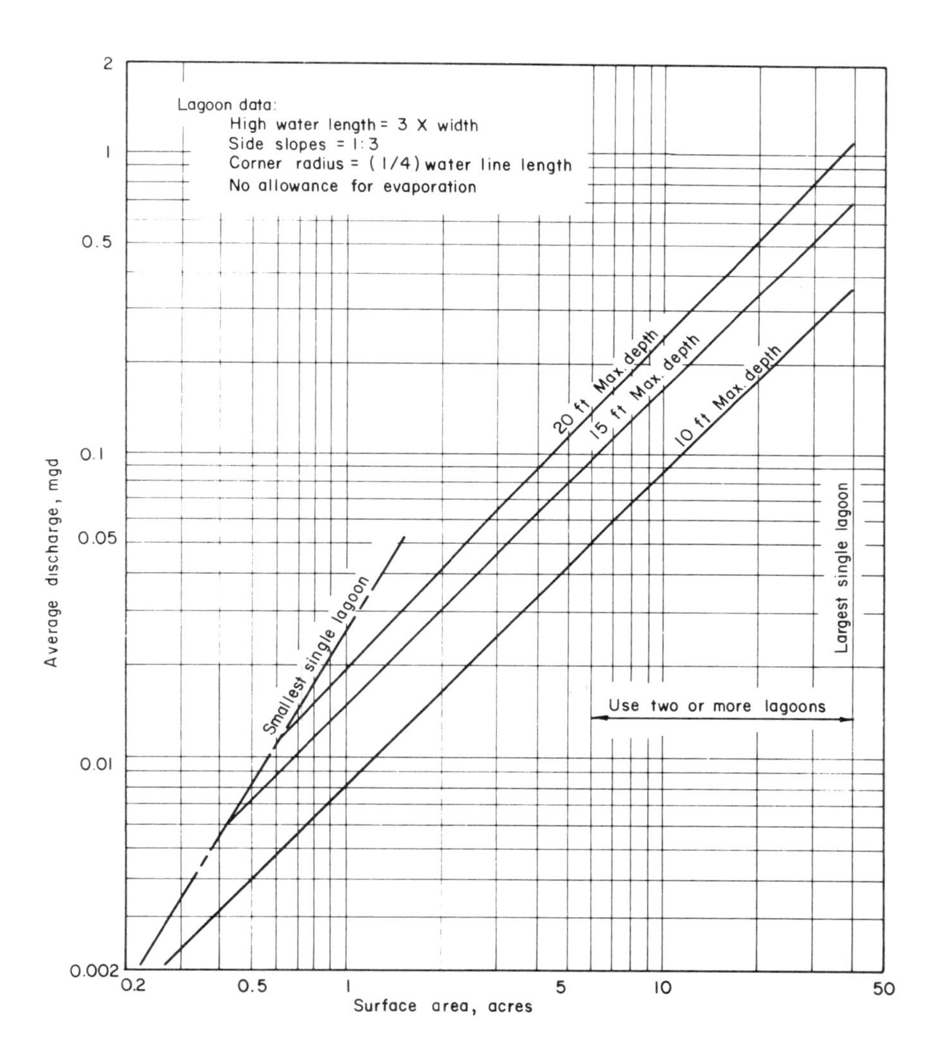

Figure 11.4. Size of single lagoon for 180-day storage and 5-ft minimum depth.

groundwater. The least expensive satisfactory lining is 15 mil vinyl protected with 6 inches (15 cm) of earth cover. The total installed cost (in March 1974) of the vinyl and earth cover varied from about $1.75 per sq yd (0.84 m^2) for less than 1000 sq yd (836 m^2) to $1.50 per sq yd (0.84 m^2) for 1000 to 20,000 sq yd (836-16,730 m^2) to $1.30 per sq yd for 20,000 to 50,000 sq yd (16,730-41,800 m^2). The cover earth must be protected from scour caused by wave action.

Aerators

Both aeration systems included herein (as well as others of their type) will give good service if properly designed and maintained. Extreme cold has caused broken connections or frozen lines in both systems in Montana. Such troubles are easily eliminated by burying pipes at least five feet below the surface in dykes and by operating the lagoons with enough water depth to prevent any ice formation around any air line at any point. The incorporation of safety provisions for ice should have little effect on costs.

In cold climates, blowers and motors should be designed with provision for the high viscosity of cold air. Furthermore, intermittent operation to save electricity has led to frozen lines, so that during cold spells the blowers should be operating continuously.

Air-Aqua hose should be cleaned at least once every three months. Gas cylinders should be housed in a shed that can be heated. Some operators are reluctant to use hydrogen chloride because of its toxicity and odor. There would be much better maintenance and hence few or no problems if operators were trained to use hydrogen chloride properly and safely. It costs about $6-7 annually per blower horsepower for the hydrogen chloride gas used in quarterly cleanings, and the operator time for manual cleaning varies from about 24 man-hours per annum for small systems (less than 2 blower hp) to perhaps as much as 140 man-hours per annum for large systems (100 to 200 blower hp). The Hinde Engineering Company[a] is evaluating several systems for fully automatic gas cleaning activated once per month by a time clock. Such a system appears to use less gas than quarterly cleaning, keeps the aerators at peak efficiency at all times, saves operator time, and would avoid the problem of reluctance of operators to use the gas.

The costs (installed and ready to operate) and the power requirements for Air-Aqua systems are given in Figures 11.5 and 11.6, respectively, and those for Polcon aerators are given in Figures 11.7 and 11.8. The costs are based on a preaeration lagoon plus aerated storage lagoons. These cost

---

[a]Hinde Engineering Co., 654 Deerfield Road, Highland Park, IL 60035.

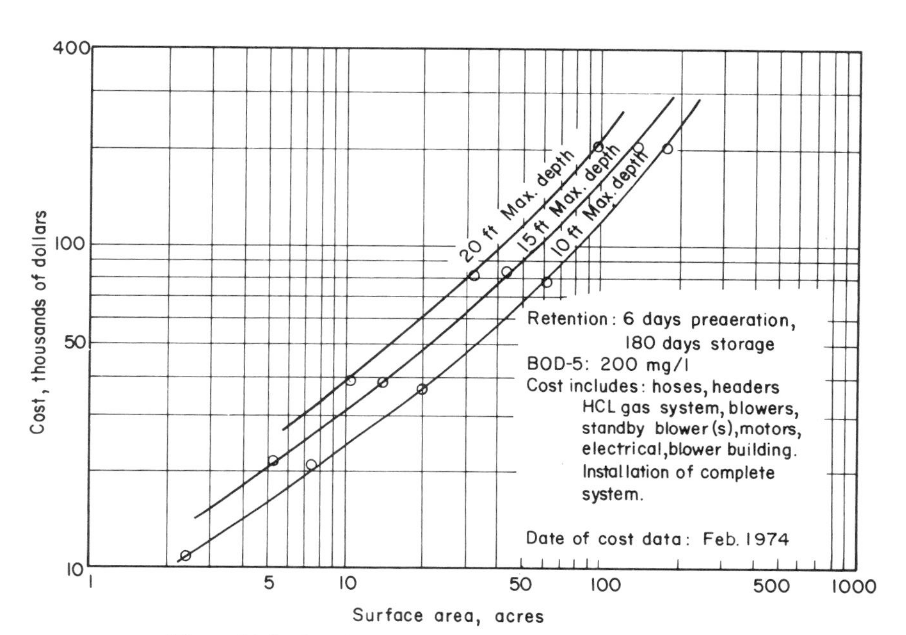

Figure 11.5. Installed cost of Air-Aqua aeration system.

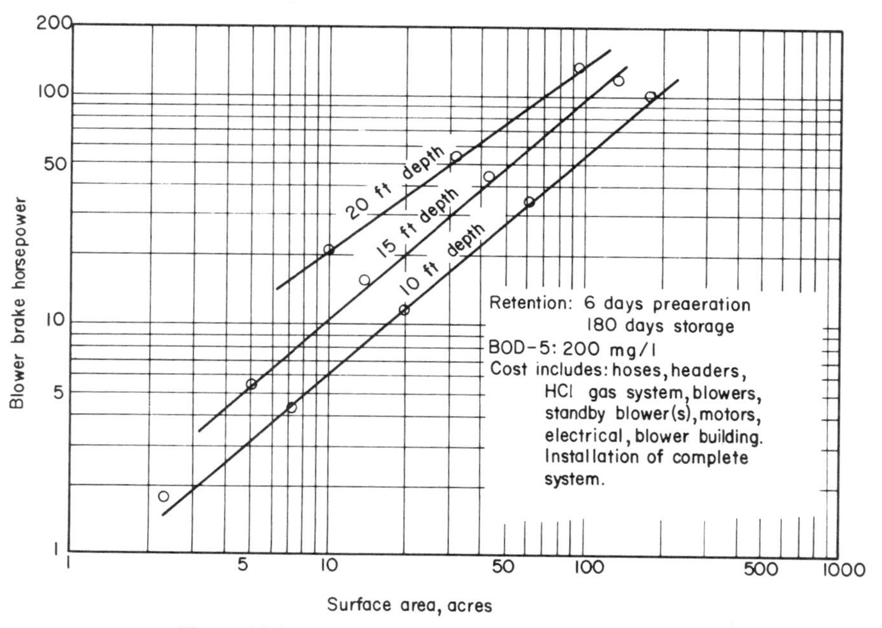

Figure 11.6. Power required for Air-Aqua blowers.

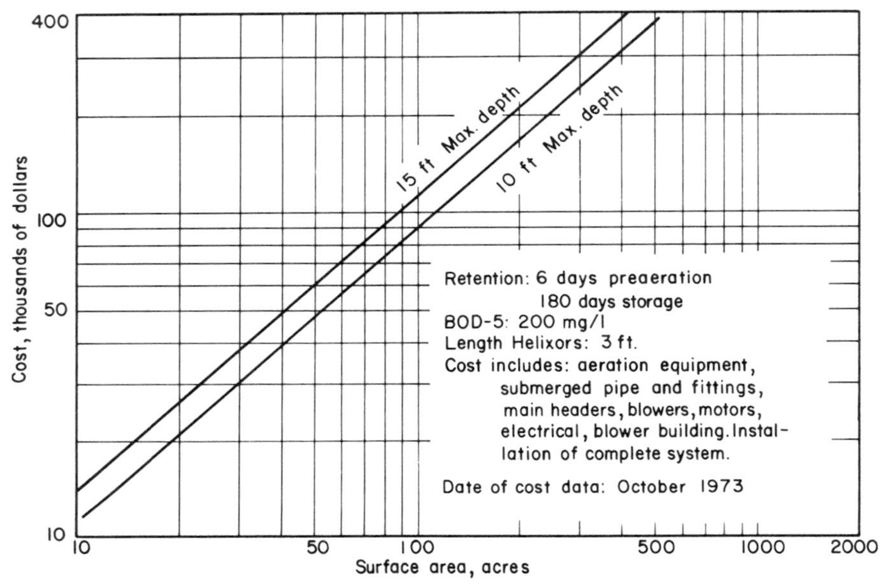

Figure 11.7. Installed cost of Polcon aeration system.

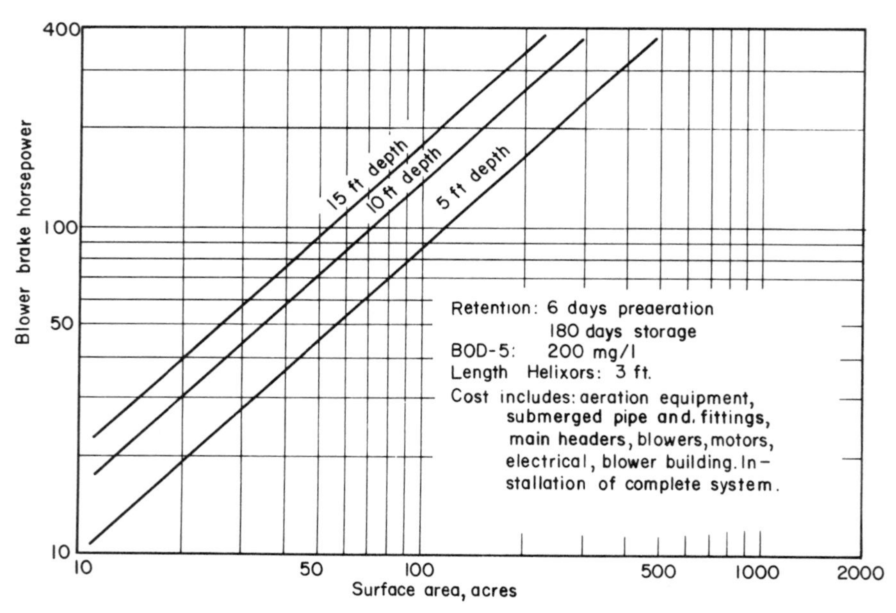

Figure 11.8. Power required for Polcon blowers.

curves cannot be used to compare costs of Air-Aqua versus Polcon systems. They are intended only for preliminary cost estimating and they are not based upon the same performance specifications. The Air-Aqua system is designed to remove about 90% of BOD-5, supply a dissolved oxygen concentration of 4 to 6 mg/l, and provide for complete turnover in 100 minutes or less. The Polcon system was designed to supply a dissolved oxygen content of about 2 mg/l. Valid comparisons can only be made on the basis of bids for equivalent performance specifications for a specific job together with a comparison of annual costs that include amortization, life, maintenance, and operation.

For both systems, the costs include all aeration equipment, all submerged piping and fittings, blowers, motors, electrical equipment, a blower building and the installation of the complete system assuming 6 days detention time in a preaeration basin plus 180 days of storage capacity for wastewater containing 200 mg/l of BOD-5. The cost of the Air-Aqua system also includes the hydrogen chloride gas cleaning system and enough gas for one cleaning operation. More data on Air-Aqua systems are given in Table 11.9.

Spray Irrigation Systems

Refer to Norum in Chapter 12 for a discussion of the costs of spray irrigation systems. Capital cost, maintenance, and operation of several different systems such as solid set, center pivot, and traveling big gun should be carefully compared on the basis of amount of use and the inherent advantages and disadvantages for a particular site and size of project.

Pumping Stations and Outfalls

The cost of pumping stations can be obtained from Patterson and Banker[30] but cost data for outfalls varies so greatly that appropriate contractors should be consulted. For example, cost data on pipe and trenches of the same depth and size vary by a factor of five or more depending on location, soil conditions, backfill requirements, and the market. Fortunately, consulting engineers are usually familiar with costs of pipelines and their installation.

Corrosion of outfall pipelines may prove troublesome, and consequently designers should determine such corrosive properties of wastewater as Langelier Index[31] and chloride concentration. The specific corrosion resistance of pipes can be obtained from pipe manufacturers.

Checklist

Willems (Chapter 1) has included a checklist that should prove useful to designers. Statements on the costs of land application systems have been prepared by Pound, Crites, and Griffes.[32,33]

Table 11.9. Air-Aqua Oxidation Systems

Model	Approximate Population	Flow mgd	lbs BOD-5 per day	Wire hp & Pressure		Lagoons			Turnover Time, min
				Low Water	High Water	No.	Bottom Size ft	Surface Acres	
	Depth water, 2 to 10 ft			**4 psi**	**7 psi**				
E10-2	300	0.03	51	1.15	1.8	2	125x243	1.29	107
E10-5	1000	0.1	170	2.3	4.3	2	250x488	3.90	91
E10-15	3000	0.3	510	5.4	11.5	2	250x1537	11.36	91
E10-40	10000	1.0	1700	20.2	34.0	2	750x1873	35.94	93
E10-120	30000	3.0	5100	60.0	102.0	4	1250x1714	53.35	100
	Depth water, 3 to 15 ft			**4 psi**	**9 psi**				
E15-7.5	1000	0.1	170	2.7	5.5	2	250x274	2.84	89
E15-15	3000	0.3	510	8.8	15	2	250x932	7.98	91
E15-50	10000	1.0	1700	21.0	45	2	250x3235	25.95	91
E15-120	30000	3.0	5100	64.0	117	2	750x3685	72.64	100
	Depth water, 4 to 20 ft			**5 psi**	**11 psi**				
E2-25	3000	0.3	510	10.0	21	2	250x625	6.33	89
E2-60	10000	1.0	1700	22.5	55	2	250x2267	20.28	91
E2-150	30000	3.0	5100	65.0	135	2	500x3861	56.66	99

All data furnished by Hinde Engineering Company based on 6 days preaeration, 180 days storage, 85-90% BOD reduction and 4 to 6 mg/l DO in effluent.

CONCLUSIONS

Land disposal of wastewater effluents is by no means new. It has been practiced for centuries, sometimes successfully and sometimes with disastrous results. At the present time it is enjoying a popular revival, and indeed under proper conditions it is a good method. But it is by no means a panacea. It is simply one more engineering alternative, and the decision to use it should be based upon consideration of all alternatives based on both economic and intangible factors.

Designers owe it to themselves and to their clients to study both pros and cons of land treatment and disposal. In this respect, unfortunately, most of the literature is heavily biased in favor of land application, so that one should be particularly careful to read contrary opinions, such as those expressed by Lee in Chapter 9. If this is done conscientiously, failures and serious problems that have plagued land application of wastewaters in the past can be avoided.

REFERENCES

1. *Water Quality Criteria.* Report of the National Technical Advisory Committee to the Secretary of the Interior, Federal Water Pollution Control Administration (Washington, D.C.: U.S. Government Printing Office, 1968).
2. "Diagnosis and Improvement of Saline and Alkali Soils," in *Agriculture Handbook No. 60.* (Washington, D.C.: U.S. Dept. Agriculture, 1954).
3. Eaton, F. M. and L. V. Wilcox. *The Behavior of Boron in Soils.* USDA Technical Bulletin 696 (1939).
4. Healy, K. A. and R. Laak. *Factors Affecting the Percolation Test,* CE 72-51 (Storrs, Conn.: School of Engineering, University of Connecticut, 1972).
5. Winneberger, J. T. and J. W. Klock. *Current and Recommended Practices for Subsurface Wastewater Disposal Systems in Arizona.* (Tempe, Arizona: Engineering Research Center, College of Engineering Sciences, Arizona State University, 1973).
6. "Standard Method for Particle-Size Analysis for Soils, ASTM Designation D 422-63," *1970 Annual Book of ASTM Standards, Part 11* (Philadelphia, Pennsylvania: American Society for Testing and Materials, 1970).
7. *Standard Methods for the Examination of Water and Wastewater,* 13th ed. (New York: American Public Health Association, 1971).
8. Pierce, D. M. "Michigan's Experience with the Ten States Guidelines for Land Disposal of Wastewater," in *Recycling Treated Municipal Wastewater and Sludge Through Forest and Cropland,* W. E. Sopper and L. T. Kardos, Eds. (University Park, Pa.: The Pennsylvania State University Press, 1973).
9. Rhindress, R. C. "Spray Irrigation–The Regulatory Agency View," in *Recycling Treated Municipal Wastewater and Sludge Through Forest and Cropland,* W. E. Sopper and L. T. Kardos, Eds. (University Park, Pa.: The Pennsylvania State University Press, 1973).

10. Blaney, H. F., L. F. Rich, W. D. Criddle, G. B. Gleason, and R. L. Lowry. "Consumptive Use of Water," *Trans. Amer. Soc. Civil Eng.* **117**, 948 (1952).
11. Pound, C. E. and R. W. Crites. *Wastewater Treatment and Reuse by Land Application Volume I – Summary,* EPA-660/2-73-006a, U.S. Environmental Protection Agency. (Washington, D.C.: U.S. Government Printing Office, 1973).
12. Romaine, J. D. "Consider Plant Food Content of Your Crop," Reprint EE-12–57 (Washington, D.C.: American Potash Institute, Inc., 1957).
13. Committee of the Great Lakes–Upper Mississippi River Board of State Sanitary Engineers. *Recommended Standards for Sewage Works.* (Albany, N.Y.: Health Education Service, 1971).
14. Pair, C. H., *et al. Sprinkler Irrigation*, 3rd ed. (Washington, D.C.: Sprinkler Irrigation Assoc., 1969).
15. Ward, G. D., G. D. Ward & Assoc., Portland, Oregon, private communication 1973.
16. Norum, E. W. Chief Project Engineer, Lockwood Corp., Gering, Nebraska, private communication 1973.
17. Sepp, E. "Disposal of Domestic Wastewater by Hillside Sprays," *J. Env. Eng. Div., Proc. Amer. Soc. Civil Engr.* **99**(EE2) (April 1973).
18. *Spray Irrigation Manual,* Publication No. 31, (Harrisburg, Pa.: Bureau of Water Quality Management, Pennsylvania Dept. of Environmental Resources, 1972).
19. Dotson, G. K. "Constraints to Spreading Sewage Sludge on Cropland," News of Environmental Research in Cincinnati, Technical Information Office, National Environmental Research Center (Cincinnati, Ohio: U.S. Environmental Protection Agency, 1973).
20. Willems, D. G., Chief, Water Quality Bureau, Montana Dept. of Public Health, private communication April 1974.
21. Minnesota Pollution Control Agency. "Recommended Design Criteria for Disposal of Effluent by Land Application," (May 1972).
22. Thomas, R. E., K. Jackson, and L. Penrod. *Feasibility of Overland Flow for Treatment of Raw Domestic Wastewater*, EPA-660/2-74-087 (Washington, D.C.: U.S. Government Printing Office, 1974).
23. *Evaluation of Land Application Systems.* Technical Bulletin EPA-430/9-74-015 (Washington, D.C.: U.S. Government Printing Office, 1974).
24. Pound, C. E. and R. W. Crites. *Wastewater Treatment and Reuse by Land Application–Volume I–Summary.* EPA-660/2-73-006a (Washington, D.C.: U.S. Government Printing Office, 1973).
25. Pound, C. E. and R. W. Crites. *Wastewater Treatment and Reuse by Land Application–Volume II–* EPA-660/2-73-006b (Washington, D.C.: U.S. Government Printing Office, 1973).
26. Powell, G. M. *Land Treatment of Municipal Wastewater Effluents. Design Factors. Part II,* Prepared for U.S. Environmental Protection Agency Technology Transfer Program, April 1975.
27. Seabrook, B. L. U.S. Environmental Protection Agency, Washington, D.C., private communication 1975.

28. Nesbitt, J. B. "Cost of Spray Irrigation for Wastewater Renovation," in *Recycling Treated Municipal Wastewater and Sludge Through Forest and Cropland,* W. E. Sopper and L. T. Kardos, Eds. (University Park, Pa.: The Pennsylvania State University Press, 1973).
29. Swanson, C. L. "New Wastewater Treatment Processes," *McGraw-Hill's 1972 Report on Business and the Environment,* F. Price, R. Davidson, and S. Ross, Eds. (New York: McGraw-Hill Book Co., 1972).
30. Patterson, W. L. and R. F. Banker. *Estimating Costs and Manpower Requirements for Conventional Wastewater Treatment Facilities,* Water Pollution Control Research Series 17090 DAN 10/71, U.S. Environmental Protection Agency. (Washington, D.C.: U.S. Government Printing Office, 1971).
31. Langelier, W. F. "The Analytical Control of Anticorrosion Water Treatment," *J. Amer. Water Works Assoc.* **28**, 1500 (1936).
32. Pound, C. E., R. W. Crites and D. A. Griffes. *Evaluation of Land Application Systems,* EPA-430/9-74-015 (Washington, D.C.: U.S. Environmental Protection Agency, 1974).
33. Pound, C. E., R. W. Crites, and D. A. Griffes. *Cost of Wastewater Treatment by Land Application,* EPA-430/9-75-003 (Washington, D.C.: U.S. Environmental Protection Agency, 1975).

SELECTED READING LIST

Annual Technical Conference Proceedings. Sprinkler Irrigation Association, Silver Spring, Md (February 1973).

Barnarde, M. A. "Land Disposal and Sewage Effluent: Appraisal of Health Effects of Pathogenic Organisms," *J. Amer. Water Works Assoc.* **65**, 432 (1973).

Culp, G. *Design Seminar for Land Treatment of Municipal Wastewater Effluents,* prepared for U.S. Environmental Protection Agency Technology Transfer Program, Culp/Wesner/Culp, Clean Water Consultants, El Dorado Hills, California.

Culp, G. "Example Comparisons of Land Treatment and Advanced Waste Treatment," Design Seminar for Land Treatment of Municipal Wastewater Effluents, Technology Transfer (Washington, D.C.: U.S. Environmental Protection Agency).

Dalton, F. E. and R. R. Murphy. "Land Disposal IV: Reclamation and Recycle," *J. Water Poll. Control Fed.* **45**, 1489 (1973).

Davis, W. K. "Land Disposal III: Land Use Planning," *J. Water Poll. Control Fed.* **45**, 1458 (1973).

Demirjian, Y. A. "Muskegon County Wastewater Management System," Design Seminar for Land Treatment of Municipal Wastewater Effluents, Technology Transfer. (Washington, D.C.: U.S. Environmental Protection Agency).

D'Itri, F., *et al.* "An Overview of Four Selected Facilities that Apply Municipal Wastewater to Land," Design Seminar for Land Treatment of Municipal Wastewater Effluents, Technology Transfer. (Washington: D.C.: U.S. Environmental Protection Agency).

Egeland, D. R. "Land Disposal I: A Giant Step Backward," *J. Water Poll. Control Fed.* **45**, 1465 (1973).

Green Land-Clean Streams: The Beneficial Use of Wastewater Through Land Treatment, Center for the Study for Federalism, Temple University, Philadelphia (1972).

Healy, K. A. and R. Laak. "Factors Affecting the Percolation Test," *J. Water Poll. Control Fed.* **45**, 1508 (1973).

Klein, S. A., D. Jenkins, R. J. Wagenet, J. W. Bigtar, and M.-S. Yang. *An Evaluation of the Accumulation, Translocation, and Degradation of Pesticides at Land, Wastewater Disposal Sites,* Contract No. USA-DADA-17-73-C-3109, Sanitary Engineering Research Laboratory, University of California, Berkeley (1974).

*Methods of Soil Analysis, Agronomy No. 9, **Part** 1 and 2,* (Madison, Wis.: American Society of Agronomy, 1965).

Pair, C. H., *et al. Sprinkler Irrigation*, 3rd ed. (and Supplement). (Washington, D.C.: Sprinkler Irrigation Association, 1969).

Pound, E. C. and R. W. Crites. "Design Factors, Part I," Design Seminar for Land Treatment of Municipal Wastewater Effluents, Technology Transfer (Washington, D.C.: U.S. Environmental Protection Agency).

Powell, G. M. "Design Factors, Part II," Design Seminar for Land Treatment of Municipal Wastewater Effluents, Technology Transfer (Washington, D.C.: U.S. Environmental Protection Agency).

Robert S. Kerr Environmental Research Laboratory. *Land Application of Sewage Effluents and Sludges: Selected Abstracts.* EPA-660/2-74-042. (Washington, D.C.: U.S. Environmental Protection Agency, 1974).

Sopper, W. E. and L. T. Kardos, Eds. *Recycling Treated Municipal Wastewater and Sludge Through Forest and Cropland.* (University Park, Pa.: The Pennsylvania State University Press, 1973).

Thomas, R. E. "Land Disposal II: An Overview of Treatment Methods," *J. Water Poll. Control Fed.* **45**, 1476 (1973).

12

DESIGN AND OPERATION OF SPRAY IRRIGATION FACILITIES

Edward M. Norum

Chief Product Engineer
Lockwood Corporation
Gering, Nebraska 69341

GENERAL PROJECT CONSIDERATIONS

The design of spray irrigation facilities for wastewater involves a conventional irrigation engineering approach with some significant extensions and changes in philosophy. In order to understand the differences it is important to review basic areas of concern to the irrigation system designer. Spraying of irrigation and wastewater involves a multidisciplinary approach. Required project background information is shown in Table 12.1. Specific design data are derived from each area; for example:

Soils

1. intake rate allowable as it relates to probable method of application, in./hr (cm/hr)
2. root zone storage as it relates to soil properties and rooting depth for probable crops to be grown, in. (cm)

Water Supply

1. quantity characterized as total flow and variability throughout growing season, mgd (m^3/day)
2. quality as it relates to possible deterioration effect on system components
3. quality as it relates to salinity, sodium, and boron content and possible leaching requirements for specific crops, mg/l

Climate

1. growing season for probable crops, days
2. rainfall as it correlates with crop water requirements in a moisture balance to develop a deficit required to be applied through irrigation, in./day (cm/day) or total in. for growing season

Conservation Needs

1. runoff control as it relates to soil loss and nuisance silting of irrigation facilities
2. soil drainage as it relates to efficient crop production through root zone aeration
3. flood protection to avoid crop loss and destruction of irrigation facilities
4. soil management to avoid top soil loss through wind and water erosion.

Table 12.1. Project Background Information and Probable Source

Item	Data Source
Site topographical features	Independent surveyor; U.S. Geological Survey Quadrangle maps
Soils: intake characteristics, root zone storage	U.S. Soil Conservation Service; county agents; extension specialists; state universities
Water supply: quantity, quality	U.S. Geological Survey; State "Dept. of Land & Natural Resources"; U.S. Dept. of Agriculture; consulting geologist
Climate: growing season, rainfall, temperature, humidity, wind	U.S. Weather Bureau
Crop: water requirements	County agents; extension specialists; state universities; consulting agronomist
Soil and water conservation needs: runoff control, soil drainage, flood protection, soil management	U.S. Soil Conservation Service; Corps of Engineers; county agents; extension specialists; state universities
Power facilities: type, location, reliability	Power company; personal survey

The design of a wastewater spray irrigation system involves these basic considerations plus others that develop from significant differences in philosophy and conditions as follows:

1. The primary objective of the system is to treat wastewater and not to maximize crop production.
2. The system will probably be managed by people with a limited background and interest in agricultural production.
3. The system must function automatically with a very high degree of reliability.
4. Federal, state, and local regulations of the system will be severe.
5. Public input to individual projects will be significant.

The importance of these differences can be understood further by reviewing the additional background data considerations:

Soils

1. The ability of soils and associated microorganisms to handle the wide range of pollutants present in the wastewater must be determined. Some pollutants are converted to usable plant nutrients, *e.g.*, the nitrification of ammonia by Nitrosomonas and Nitrobacter.[1] Other pollutants, such as heavy metals, are held on the soil's clay complex.[2] In humid areas, the limiting factor in system applications is the nutrient loading rate which best balances the crop requirements. If the cover crop has no definable requirement for the pollutant, levels of buildup in the root zone must be monitored to avoid toxic buildups. This can also mean cation exchange capacity determinations with arbitrary limitations on the percentage of saturation allowable.
2. The ability of the soil to filter the wastewater solids and the effect on intake rate must be known. This can range from no effect to complete plugging. If plugging is possible the importance of resting the site between applications must be studied as a means of reestablishing a suitable intake rate.
3. The influence of temperature on the intake rate must be studied if spraying is to occur beyond the normal growing season. Petersen[3] noted 1/3 to 1/2 reduction in hydraulic conductivity with a drop in temperature of 15°F (8.3°C).

Water Supply

1. The quantity must be characterized for the whole year to allow for storage when spraying is not possible.
2. The quality must be studied to allow characterization of all pollutants. The state of Pennsylvania,[4] for example, requires an analysis of the following items:

BOD and COD	chromium	phosphorus
total iron	sulfate	alkalinity
manganese	chloride	MBAS
aluminum	fluoride	TSS
copper	nitrogen (Kjeldahl)	pH
zinc	nitrogen (ammonium)	temperature
nickel	nitrogen (nitrate)	

Climate

1. The relationship of rainfall to storm intensity and duration, which in turn relates to runoff hazard. If a significant runoff hazard exists, regulation may require containment, sampling, and retreatment.
2. Wind is related to the spreading of odors and disease organisms in aerosols. Wind data may be used to modify operating procedures to avoid spraying during high wind periods. Also, system buffer zone and windbreak requirements may relate to wind speed and direction.

Crop

1. In addition to water requirements, the crop's ability to tolerate high levels of root zone moisture must be studied. As noted by Sopper,[5] this accounts for the popularity of reed canarygrass on disposal sites.
2. The ability of crops to take up various elements[6] must be determined as it relates to allowable loading rates. Crops may be selected or bred specifically for their ability to cycle or tolerate high levels of particular elements.

Conservation Needs

1. Runoff control becomes an essential part of the system design. Runoff could pollute adjacent streams and subject the system owner to penalties.
2. Drainage systems can be an integral part of the treatment process where the objective is to renovate wastewater and recycle it for other uses.
3. Flood protection is essential to ensure the continuous functioning of the system and avoid stream pollution.
4. Soil management is also essential to ensure that the pollutants remain on the disposal site.

In addition to these extended conventional considerations, the system designer must deal with input from several new areas. In the first place, he must understand the type of pretreatment of the wastewater and the resulting effluent characteristics. Solids content relates to potential nozzle and valve plugging. Chemical characteristics relate to allowable loading rates and possible deteriorating effects on system components. Toxicity relates to potential odor and nuisance problems and detrimental effects on crops. Flow rate and variability relate to required system capacity, and disinfection method and adequacy relate to aerosol drift and resulting buffer zone requirements.

The system designer must also understand the site hydrogeology, the potential for groundwater contamination, the possible presence of unfavorable geological formations such as limestones and dolomites that could develop into sink holes, the movement of the groundwater away from the disposal site, and the depth to groundwater.

In addition, the system designer must be familiar in detail with the wide assortment of local, state and federal regulations that will affect the design. The primary motivation for the laws stemmed originally from a concern for public health. More recently, Public Law 92-500 expressed the will of the people to improve the quality of life. In some cases, more nebulous concerns motivated the development of regulations dealing with vested interests in land development, odor and nuisance problems, and vested interests in other methods of pollution control.

Eventually the technical research available will be extensive enough to provide firm system design requirements. This will help the designer by narrowing the alternatives and establishing important parameters such as the amount of buffer zone required, the minimum depth to water table, loading rates, exchange capacity saturation, pretreatment, disinfection, leachate quality, spray period, uses for harvestable cover crops, runoff water quality, and the level of system reliability.

GENERAL EQUIPMENT CONSIDERATIONS

There has been a tendency to apply irrigation technology directly to the design of systems for the spray irrigation of wastewater. This has sometimes led to poorly designed systems and even to complete failures. This tendency can be explained in part by the inexperience of the system designers and in part by the convenience of using readily available hardware. Very real differences exist between the design of irrigation systems and the design of systems for the spraying of wastewater. Some of these differences are: application uniformity, application rate, climatic influences, and solids handling capabilities.[7]

Historically, the uniformity of sprinkler systems has been measured and characterized by procedures first developed by Christiansen.[8] This resulted in the widespread use of a measure of application uniformity called "Christiansen's Uniformity Coefficient" (UC) as first defined by Christiansen and refined for specific systems by others.[9,10] UC values commonly used in irrigation design range from 70-95%. The lower values are appropriate for lower valued crops in areas of low-cost water. The higher values are used on high-valued crops irrigated with high-cost water. It has been suggested[10] that a value of 80% or better is appropriate for center pivot systems.

A modified method of characterizing uniformity has been developed by Norum.[11,12] The significance of the procedure can best be understood by referring to Figure 12.1, which shows application data for a commonly used solid set system.

If the design application is 2.0 inches (5.08 cm), it can be seen from Figure 12.1 that about 50% of the disposal site will receive an over-application ranging up to 3.75 in. (9.52 cm). If the renovating capability of the soil profile has been accurately determined to be 2.0 in. (5.08 cm), then the quality of renovation has been seriously jeopardized in the over-applied

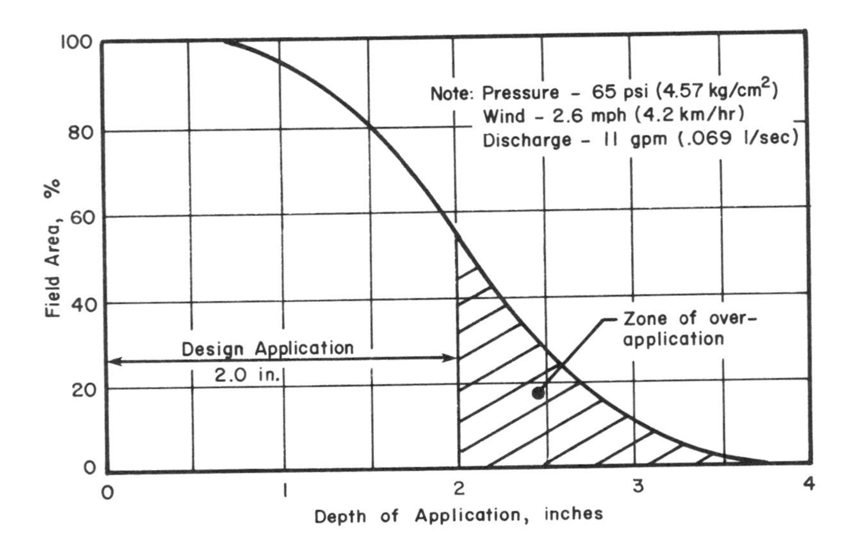

Figure 12.1. Application uniformity data for a 60 ft x 80 ft solid set system.

areas. This is especially significant if the quality of renovation is to approach drinking water standards. If the design application is based on a nutrient constraint of 200 lb of nitrogen per acre (224 kg per ha) per year, for example, then an actual range of from about 75 to 375 lb per acre (84 to 420 kg per ha) is possible.

For a given system design many factors affect the actual composite uniformity probable on a disposal site: soil lateral permeability, presence of restricted permeability layers in the soil profile, movement of surface water during application, wind velocity and direction, and operating options. The type of uniformity of application curve to be expected is unique to the system design and site conditions and can only be determined by test. The results of actual tests on equipment commonly used for solid set disposal systems are shown in Figure 12.2. The abcissa is the actual application divided by the mean application following a method outlined by Norum.[11] The data in Figure 12.2 help to determine the spacing that should be used. From the curves, the deterioration in uniformity of application is apparent when the same sprinkler is used at increased spacings.

For the spraying of wastewaters, applications to the disposal site should be limited to the renovating capability of the soil. Following a method suggested by Norum,[12] and using the data in Figure 12.2, the design

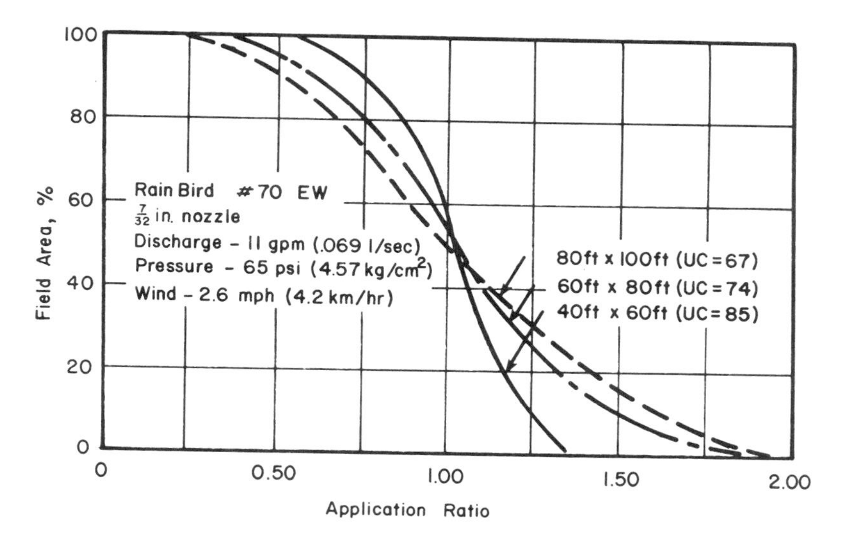

Figure 12.2. Uniformity of application curves for a typical solid set system.

applications for the indicated spacings to comply with a 2.0 in. (5.08 cm) per week constraint over 90% of the area are:

Application, in. = 2.0 ÷ application ratio at 10% field area		
40 ft x 60 ft	1.63 in.	2.0/1.23 = 1.63 in.
60 ft x 80 ft	1.32 in.	2.0/1.52 = 1.32 in.
80 ft x 100 ft	1.20 in.	2.0/1.67 = 1.20 in.

These data can be used in framing a disposal area versus system cost tradeoff. The 40 ft x 60 ft (12.2 m x 18.3 m) spacing achieves good uniformity, and uses a minimum of land, but has a high equipment capital cost. The 80 ft x 100 ft (24.4 m x 30.5 m) spacing has poor uniformity, fails to completely charge the root zone and requires more land, but has a lower equipment capital cost. Ultimately the decision on system uniformity must be made on the basis of this kind of analysis. It varies significantly from irrigation procedures and cannot be solved by rules of thumb or generalizations. The consequences of inadequate design are potentially more serious than with irrigation system design.

Spray irrigation systems for wastewater are generally subjected to far more sensitive conditions of soil intake rate than crop irrigation systems.

---

*Rainbird Sprinkler Mfg. Corp., Glendora, California 91740.

In addition, the consequences of a poor match between system application rate and soil intake rate are potentially more serious. Runoff can pollute nearby streams or accumulate in hollows and kill the cover crop. Also nuisance, odor, and fly problems can develop, either of which can force a project to shut down.

Irrigation generally proceeds when the root zone is dried out and the crop requires moisture, thereby putting intake rates at a maximum. With the spraying of wastewaters, however, the soil profile is frequently near saturation, particularly in the humid eastern part of the country and during slower growing times of the year. Under these conditions, the intake rate of the soil is greatly reduced and approaches the basic saturated conductivity. This reduction in intake rate can best be studied by an example soil shown in Figure 12.3.

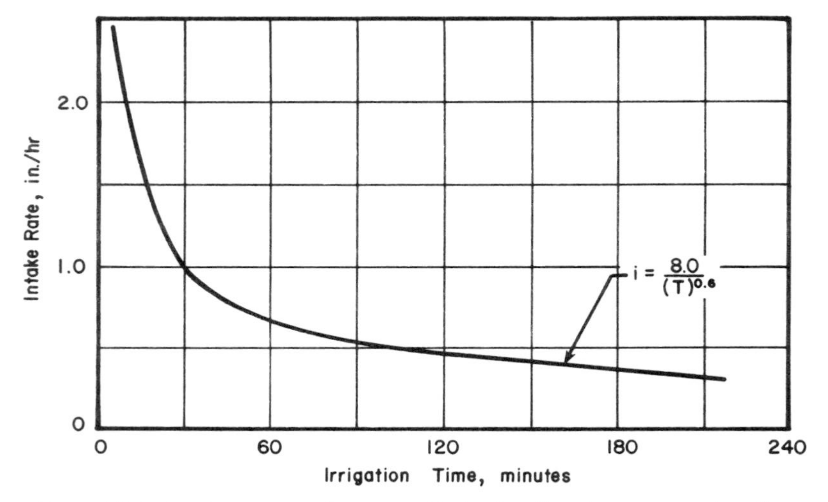

Figure 12.3. Intake rate as a function of irrigation time for an example soil. (After Norum.[12])

Irrigation systems with application rates of 0.50 to 0.75 in. (1.27 to 1.91 cm) per hr could probably be used successfully on this soil. As a wastewater disposal site, however, an application rate would have to be reduced to 0.20 to 0.30 in. (0.51 to 0.76 cm) per hr to avoid runoff. This example deals only with the effect of antecedent moisture content. It should be noted that several other factors can also affect intake rate, particularly: cover crop, surface detention, soil clogging by effluent particles, breakdown of the soil structure by chemical or mechanical processes, and temperature effect on water properties.

Because of the difficulty of adequately predicting the composite effect of these variables, the best system design would be based on actual field determination of intake rate. This determination can be made with ring or sprinkler infiltrometers. After definitive information is available on the actual soil intake rate, it remains to match this soil property to the measured performance of the irrigation system. This can best be explained using actual data for typical systems, as shown in Figure 12.4.

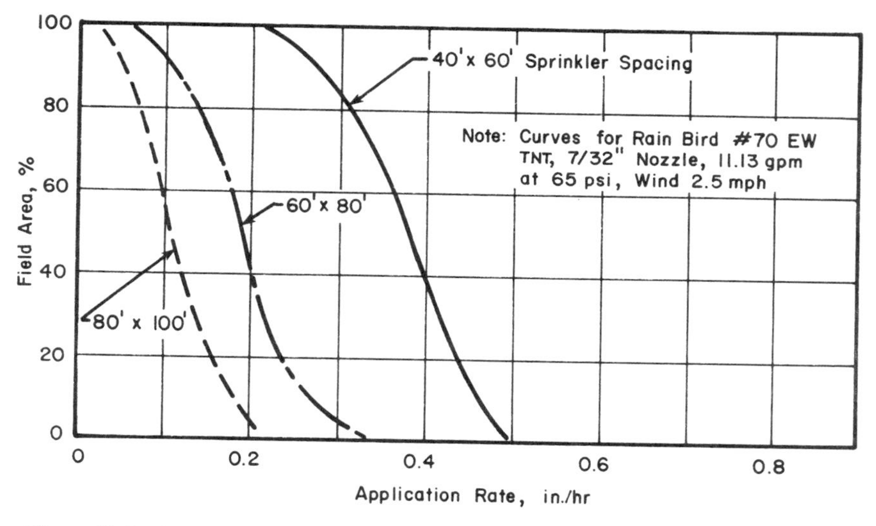

Figure 12.4. Application rate curves for a typical solid set system. (After Norum.[12])

Assuming that the system is to distribute wastewater occasionally under saturated soil conditions (as characterized by Figure 12.3), the 40 ft x 60 ft (12.2 m x 18.3 m) spacing would result in runoff over 80% of the area and would be unacceptable. The 60 ft x 80 ft (18.3 m x 24.4 m) spacing would have runoff on about 5% of the area, which probably would be acceptable. No runoff would be expected on the 80 ft x 100 ft (24.4 m x 30.5 m) spacing, but the uniformity of application as noted in Figure 12.2 would be badly deteriorated. Good practice requires a careful determination of soil intake rate as noted in Figure 12.3 and a careful matching with system characteristics as shown in Figure 12.4.

A major climatic influence is the effect of wind on aerosol drift. Relatively little is known about the behavior of these aerosols and in particular the potential disease hazard when treated and sprayed wastewater aerosols drift off the disposal site. A convincing argument can be made for the deposition of disease organisms on the soil where they are held until they die as opposed to discharge into receiving waters.

In an effort to minimize the aerosol hazard at Muskegon, the machines utilized spray bars mounted about 9 ft (2.74 m) above ground level and operated at 10 psi (0.703 kg/cm^2). An empirical calculation of drift indicated that this would assure that all droplets larger than 150 microns in diameter would fall within the 200-ft (61-m) buffer zone in winds up to 20 mph (32.2 km/hr). The estimated median particle diameters for the nozzles selected ranged from 800 to 1800 microns. The adequacy of this design was verified by an actual field measurement of aerosol drift.

It seems apparent that high-pressure sprinklers will be very susceptible to aerosol drift and should be studied before final project designs are made. Traveler systems are potentially the most hazardous, with large volumes of water discharged at high pressures of 85 to 100 psi (5.98 to 7.03 kg/cm^2).

For protection against the spread of diseases by aerosol drift, state regulatory agencies require buffer zones around disposal sites ranging in width from 200 ft (61 m) to 1/4 mile (0.4 km). In addition, system designers would be wise to incorporate windbreak trees into the buffer zone and even to consider shutting down during periods of unusually high wind. Careful attention should also be given to adequate disinfection and careful monitoring of the operation.

The spray equipment must have the capability of passing the maximum particle size present in the effluent. A particle size determination must be made and the orifices sized accordingly. If this conflicts with application rate and uniformity objectives, straining equipment is required. Sprinkler orifices vary from 3/32 in. (0.24 cm) to almost 2.0 inch (5.1 cm). Nozzles for spray bars vary to a minimum of 0.09 in. (0.23 cm) in diameter, which requires careful study of screening requirements.

The difficulty of orifice plugging is accentuated when, for example, sludge is to be sprayed on relatively tight soils. Traveler systems with large orifices have been successfully used but present the inherent problem of aerosol drift. Some effort is being made to adapt low pressure orifices with drop hoses to the application of sludge to growing crops. More subtle plugging problems can occur when fats, hair, and fine soil particles accumulate and eventually produce a plug.

CHARACTERISTICS OF SPECIFIC SYSTEMS

An understanding of the capabilities of the commercially available spray disposal systems is important to the system designer. For the most part, the hardware was developed by the irrigation industry and is currently applied to an aggregate total of over 10,000,000 acres (4,050,000 ha). Using irrigation systems with minor modifications ensures the availability of equipment and repair parts generally mass-produced at realistic prices.

The five most popular systems suitable for spray disposal are described in Table 12.2, in which emphasis is placed on the characteristics and requirements unique to spraying wastewater.

Portable Solid Set

A portable system indicates that all system components are lightweight, quick-coupled, and easily moved by hand. Solid set refers to a layout configuration with a sprinkler located at each intersection of a regular grid system. A sketch of a typical system is shown in Figure 12.5. The main lines and sub mains are usually 4-, 6-, or 8-inch (10.2, 15.2, or 20.3 cm) diameter aluminum tubing in 20 or 30 foot (6.1 or 9.1 m) lengths. The laterals are usually 2-, 3-, or 4-inch (5.1, 7.6, or 10.2 cm) aluminum tubing in 30 or 40 foot (9.1 or 12.2 m) lengths with sprinklers attached at one end or in the middle by means of a saddle. The aluminum tubing (made to ASAE[13] specifications) is class 150, which requires a minimum 450 psi (31.6 kg/cm^2) hydrostatic test and allows for a maximum operating pressure of 150 psi (10.5 kg/cm^2).

The development of the quick-coupled joints dates from the late 1940s when aluminum companies were looking for new peace-time markets. Catalogs are available from a number of manufacturers. Care must be taken in selecting a coupler to be sure that it will stay together during continuous operation over a range of temperatures. This can be a chronic problem with systems installed in remote areas receiving minimal inspection. Couplers with a "mainline" design are preferred. Pressure rating on couplers is usually adequate for the intended application. However, system pressures over 100 psi (7.0 kg/cm^2) should be reviewed with the coupler supplier to ensure suitability. Few couplers are designed to operate at more than 150 psi (10.5 kg/cm^2). Problems with coupler blowouts are most frequently encountered at fittings such as elbows, tees, and end plugs where additional holding force is required to resist dynamic forces. The dynamic forces can be related to water hammer caused when air is purged from the system. The magnitude of the surge pressure can be reduced by the installation of air relief valves and by the use of slow reacting control valves.

Corrosion of aluminum pipe and couplers can be a severe problem with some types of effluent. It should be studied before a system decision is made. It is especially severe at the point of contact between the pipe and the continuously moist soil. In some cases, a zinc-coated coupler will act as a sacrificial anode, thereby protecting the aluminum pipe.

Sprinkler spacings are usually 30 by 50 ft (9.1 by 15.2 m) to 40 by 60 ft (12.2 by 18.3 m) where excellent uniformity is required and 60 by 80 ft (18.3 by 24.4 m) or even 80 by 100 ft (24.4 by 30.5 m) where less uniformity is appropriate. It is good practice to limit lateral pressure loss

Table 12.2. Characteristics of Major Types of Spray Disposal Systems

Type of System	Maximum Field Slope, %	Ave. Water App. Rates in./hr		Field Shape	Field Surface Conditions	Max. Crop Height, ft	Labor Required, hrs/ac/irr	Size of Single System, acres	Approximate Cost, $/acre[a]	Wind Drift Hazard	Solids Handling Capability	Comments
		Min	Max									
Portable solid set	no limit	0.05	2.0	any shape	no limit	no limit	0.20-0.50	no limit	400-900	average	poor	Well-suited for field crop and forest disposal sites as reported by Myers,[14] and Frost, Towne and Turner[15]
Buried solid set	no limit	0.05	2.0	any shape	no limit	no limit	0.05-0.10	no limit	400-1000	average	poor	Well-suited for small systems spraying field crops where completely automatic operation is required[16]
Side wheel roll	5-10	0.10	2.0	rectangular	reasonably smooth	3-4	0.10-0.30	20-80	100-300	average	poor	Best suited for periodic spra spraying of industrial wastes (such as cannery wastes) on field crops
Traveler	no limit	0.25	1.0	rectangular	roadway for hose and sprinkler unit	no limit	0.10-0.30	40-100	120-250	high	excellent	Well-suited for spraying sludge on field crops as demonstrated by Metrol. San. Dist. of Greater Chicago, Fulton County, Ill.
Center pivot	5-15	0.005	0.090	circular	clear of high obstructions, path for towers	8-10	0.05-0.15	40-160	180-350	low	average	Well-suited for large-scale spraying of wastewater on field crops as demonstrated at Muskegon, Michigan[17,18]

[a]Does not include cost of water supply, pump, power unit and mainline. Best source of equipment and data is irrigation system vendors.

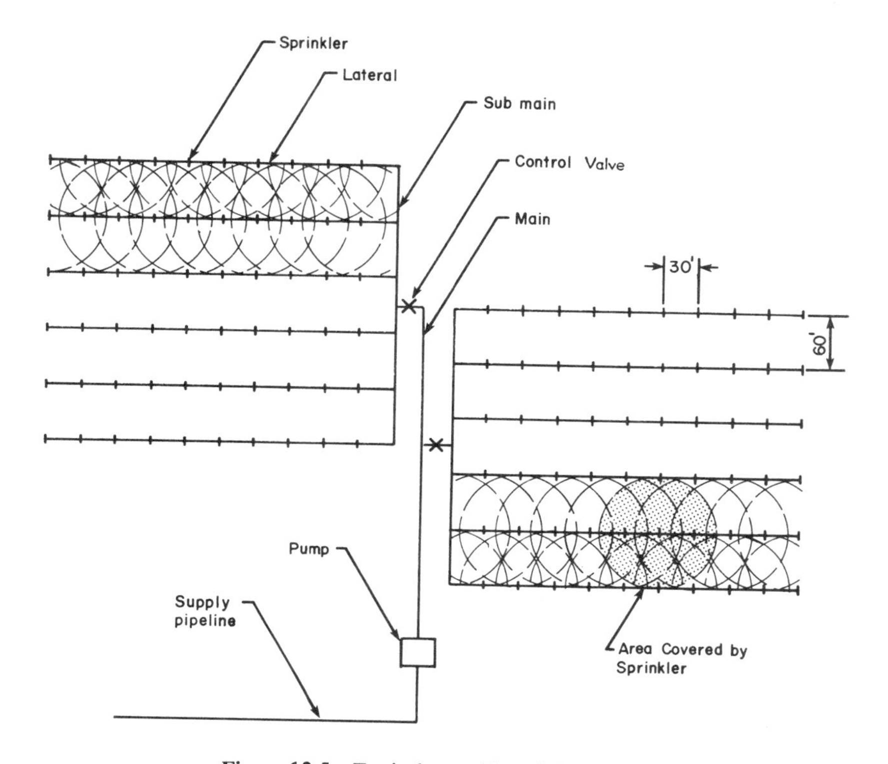

Figure 12.5. Typical portable solid set layout.

to 20% of the operating pressure. This limits variations in sprinkler discharge to about 10%. Sprinklers are available with a wide range of hydraulic and performance capabilities, as shown in Table 12.3. For spraying wastewater, sprinklers in the intermediate pressure range are usually preferred.

Since the soil profile is being relied upon to renovate the wastewater, the system should be installed with a minimum of disturbance to the soil and the cover crop. The portable solid set system is preferred in this regard, particularly for wooded sites.

Example

Additional design considerations can best be understood by a sample calculation (see also Myers[14]).

Given (1) average flow rate, 0.5 mgd (1893 m^3/day)
(2) average weekly loading rate: 2 in. (5.1 cm)

Table 12.3. Classification of Sprinklers and Their Adaptability

Type of Sprinkler	General Characteristics	Range of Wetted Diameters	Recommended Minimum Application Rate	Jet Characteristics[a]	Moisture Distribution Pattern[b]	Adaptations and Limitations
Low pressure 5-15 psi	Special thrust springs or reaction-type arms	20-50 ft	0.40 in./hr	Waterdrops large due to low pressure	Fair	Small acreages; confined to soils with intake rates exceeding 0.50 in./hr and to good ground cover on medium- to coarse-textured soils
Moderate pressure 15-30 psi	Usually single-nozzle oscillating or long-arm dual-nozzle design	60-80 ft	0.20 in./hr	Waterdrops fairly well-broken	Fair to good at upper limits of pressure range	Primarily for undertree sprinkling in orchards; can be used for field crops and vegetables
Intermediate pressure 30-60 psi	Either single or dual-nozzle design	75-120 ft	0.25 in./hr	Waterdrops well-broken over entire wetted diameter	Very good	For all field crops and most irrigable soils; well-adapted to overtree sprinkling in orchards and groves and to tobacco shades
High pressure 50-100 psi	Either single or dual-nozzle design	110-230 ft	0.50 in./hr	Waterdrops well-broken over entire wetted diameter	Good except where wind velocities exceed 4 mph	Same as for intermediate pressure sprinklers except where wind is excessive
Hydraulic or giant, 80-120 psi	One large nozzle with smaller supplemental nozzles to fill in pattern gaps; small nozzle rotates sprinkler	200-400 ft	0.65 in./hr	Waterdrops extremely well-broken	Acceptable in calm air; severely distorted by wind	Adaptable to close-growing crops that provide good ground cover; for rapid coverage and for odd-shaped areas; limited to soils with high intake rates
Undertree low-angle, 10-50 psi	Designed to keep stream trajectories below fruit and foliage by lowering the nozzle angle	40-90 ft	0.33 in./hr	Waterdrops fairly well-broken	Fairly good; diamond pattern recommended where laterals spaced more than one tree interspace	For all orchards or citrus groves; in orchards where wind will distort overtree sprinkler patterns; in orchards where available pressure is not sufficient for operation of overtree sprinklers
Perforated pipe 4-20 psi	Portable irrigation pipe with lines of small perforations in upper third of pipe perimeter	Rectangular strips 10-50 ft wide	0.50 in./hr	Waterdrops large due to low pressure	Good pattern is rectangular	For low-growing crops only; unsuitable for tall crops; limited to soils with relatively high intake rates; best adapted to small acreages of high-value crops; low operating pressure permits use of gravity or municipal supply

[a] Assuming proper pressure-nozzle size relations.

[b] Assuming proper spacing and pressure-nozzle size relations.

Assumed (1) sprinkler spacing, 30 by 60 ft (see Figure 12.5) (9.1 by 18.3 m)
(2) sprinkler: 5/32 in. nozzle, 5.22 gpm at 55 psi, wetted dia 100 ft (0.40 cm nozzle, 0.33 l/sec at 3.87 kg/cm^2, wetted dia 30.5 m)

Required (1) disposal site area, acres
(2) application rate, in./hr and application period, hr
(3) operating mode

$$\text{Disposal site area} = \frac{(0.5 \text{ mgd})\ (7 \text{ days/wk})}{(0.027154 \text{ mg/ac-in.}) \times 2 \text{ in./wk}} = 64.4 \text{ ac } (26.1 \text{ ha})^{a}$$

As shown in Figure 12.5, this could be two blocks of about 32 acres (13.0 ha) each or some other configuration that fits the topography.

$$\text{Application rate} = \frac{(5.22 \text{ gpm})\ 96.3}{(30 \text{ ft})\ (60 \text{ ft})} = 0.28 \text{ in./hr } (0.71 \text{ cm/hr})^{b}$$

$$\text{Application period} = \frac{2 \text{ in./wk}}{0.28 \text{ in./hr}} = 7.1 \text{ hr/wk}$$

Again, assuming two areas, each area would have to operate 7.1 hr/wk applying 2 in. (5.1 cm) to dispose of an average flow of 0.5 mgd (1893 m^3/day). A surge reservoir would have to be provided to store the effluent when the system was not in operation and to allow for inclement weather.

The appropriate operating mode is still being investigated. The state of Pennsylvania[4] favors spraying one day per week and allowing six days of rest between applications for the soil to drain and reaerate. This would suggest for our example a spray period on Monday of 7.1 hours and Thursday of 7.1 hours to help minimize storage requirements. In more arid regions where the objective is to maximize crop production, daily operation of about 1 hour, applying a gross application of 0.28 in. (0.71 cm) may be appropriate. The actual operating time should be selected to minimize spraying hazards such as spray drift and contact with the public. Spraying from perhaps 11:00 pm to 6:00 am would probably be best.

In laying out a portable solid set system, care must be taken to avoid trapping water in the pipe during freezing weather. All water must be drained from the system, including that which may only partly fill a pipe in a low spot. Adequate drainage can be accomplished by a combination of the following practices.

---

[a]Does not allow for buffer strip, system expansion or emergency spray area.
[b]This is an average application and ignores spray losses.

(1) Slope all lines on a minimum 2% grade either back toward the pump or toward the end of the laterals. Sloping and draining back toward the pumps is preferred because the effluent can be diverted back into the reservoir and resprayed, thereby avoiding a possible runoff pollution problem. If the system drains toward the end of the lateral, an automatic draining end plug can be used. Care must be taken to ensure that the drainage water remains on the disposal site and is absorbed by the soil. In some cases, it will be necessary to support the laterals on wooden frames in order to maintain a uniform grade.

(2) Couplers are available with a self-draining feature. Usually this involves a modification of the gasket design so that it does not seal at very low pressures. The use of self-draining gaskets is recommended because it tends to distribute pipe drainage water over all the disposal site.

(3) Inexpensive automatic self-draining valves can be installed. These are commonly used on a variety of irrigation systems and fit into a drilled hole in the pipe wall. They provide auxiliary draining capacity for larger pipes or minimum grades. To ensure adequate clearance around the drain valve or the self-draining gasket, the pipe may be set above ground level on a wooden block, or an outlet with an attached pan may be used. Care must also be used in selecting control valves so that no valve cavities exist that will trap water, thereby freezing and damaging the valve. Some valves have a built-in feature that adequately drains these cavities. In many installations, air is used as a control circuit fluid, thereby avoiding freezing problems.

Buried Solid Set

A buried solid set is a variation of the portable solid set in which all piping is buried below the frost line. Piping materials must be selected to resist the deterioration possible in buried applications. The most commonly used material is PVC. Some asbestos cement and steel pipe is used for main lines. The outstanding disadvantage of the buried solid set system is its high price. System costs in Table 12.4 were noted by Williams[16] for systems in Michigan.

Table 12.4. Costs for Buried Solid Set System. (After Williams.[16])

Project	Cost of Spray Irrigation System, $/ac
Belding	5,100
Roscommon Village	3,777
Cassopolis	1,080

There are additional disadvantages: (1) the design is fixed upon installation and cannot be easily varied, as may be required to meet changing field or system considerations, (2) the network of risers and sprinklers presents a mechanical obstruction to efficient farming operations, (3) the effluent must be well-screened to eliminate plugging problems, and (4) the site will be badly disturbed by pipeline installation equipment. The last is especially detrimental if the system is to be installed in a forest area. But in spite of the negative aspects, Williams[19] subsequently noted "We are now of a mind that we will probably not design any more spray irrigation systems using anything other than solid set systems."

The most significant advantages of the solid set system are: (1) low water application rates, (2) adaptability to any size or shape of field, (3) suitability for a wide variety of crops, and (4) adaptability to automatic operation requiring a minimum of operating labor. This can be especially important for small municipalities where full-time system supervision cannot be justified.

Wind and Ice

Both portable and buried solid set systems use intermediate pressure sprinklers that are badly affected by high winds. The wind effect is twofold: aerosol drift is increased, requiring either wider buffer zones or a system shut-down during periods of high wind; and the uniformity of application deteriorates, requiring lighter total applications to avoid overloading the soil profile in the wetter portions of the pattern. There is a need for a sprinkler especially designed to meet the needs of effluent spraying. One special problem is the build-up of ice on the sprinkler head, which eventually stops rotating. Better success has been achieved with rocker and reaction sprinkler heads.

Side Wheel Roll

Side wheel roll irrigation systems were developed in an effort to reduce the amount and type of irrigation labor required. The system consists of a lateral line of 4- or 5-inch (10.2 or 12.7 cm) aluminum tubing mounted on 5-, 6-, or 7-ft (1.5, 1.8, or 2.1 m) diameter wheels spaced 30 to 50 ft (9.1 to 15.2 m) along the pipe. A drive unit is mounted in the center of a typical 1/4 mile (0.4 km) system. Heavy soils or longer systems up to 1/2 mile (0.8 km) require additional drive units. Another recent development is the availability of automatic controls that allow the system to be moved from the water supply end.

A more recent development has a prime mover attached at one end of the system with the power transmitted through a long drive shaft. The system is moved dry with the irrigation tubing acting as an axle to transmit the torque to the wheels. The wheel size is related to the desired sprinkler spacing and the height of the crop to be grown. Usually, sprinklers are selected using the same considerations involved in solid set systems. Sprinklers are frequently spaced 30 or 40 ft (9.1 to 12.2 m) along the pipe and mounted on a self-aligning connection that allows the sprinkler to remain upright in a suitable sprinkling position.

The distance of the lateral move is related to a fixed number of wheel revolutions. For example, a 5-ft (1.5 m) wheel making 3, 4 or 5 revolutions results in a lateral move of about 50, 60, and 80 ft (15.2, 18.3, and 24.4 m) respectively. Combining this with a lateral sprinkler spacing of 40 feet (12.2 m) results in an overall sprinkler spacing of 40 x 50 ft (12.2 x 15.2 m), 40 x 60 ft (12.2 x 18.3 m), and 40 x 80 ft (12.2 x 24.4 m). For the 40 x 60 ft (12.2 x 18.3 m) spacing and a 10-gpm (0.63 l/sec) sprinkler, the average application would be:

$$\text{application rate} = \frac{(10 \text{ gpm})\ (96.3)}{(30 \text{ ft})\ (60 \text{ ft})} = 0.54 \text{ in./hr } (1.37 \text{ cm/hr})$$

Assuming a required gross application of 2 in. (5.1 cm), the system sets 4 hours in each location. After the set, the system is shut down by hand, disconnected from the water supply, moved and reconnected. Connection to the mainline is made by a flexible hose or a telescoping aluminum pipe section. Drain valves are installed along the pipe so that all water is drained when the pipe is moved. Assemblies of cartwheels can be provided to tow the system axially to another field. Also sections of the lateral can be dropped off to allow the system to move into narrower portions of the field. The system should have a braking device to resist movement in high winds.

As noted in Table 12.2, the strong point of the system is its relatively low cost. It is probably best suited for seasonal spraying of industrial wastes, such as might occur in fruit and vegetable processing applications. The system would also be well-suited to spraying a waste that has a potentially toxic effect on crops and soils requiring long rest periods between applications. In such situations the entire system could be moved periodically to a new area.

Since operation of the system requires direct personal contact, it is not well-suited to spraying sanitary wastes. Additional potential problems include: (1) a limitation on crop height of 2 to 3 ft (0.61 to 0.91 m), (2) possible corrosion problems as water is sprayed all over the equipment (whereas solid set and pivots spray directly on crops and limit the spray striking equipment), (3) mechanical failures because the units must be rolled over wet fields, and (4) the restrictions that the field must be free of obstructions and generally flat or slightly rolling.

Traveler

A traveler is a vehicle designed to carry a high-capacity sprinkler and a variable speed winch powered by an internal combustion engine or a water-powered system. The water-powered winch can be driven by a water turbine or a ratcheting cylinder drive. The unit moves itself along a travel lane at a predetermined speed, pulling a flexible hose for its water supply. The hoses are usually 4 to 4½ inches (10.2 to 11.4 cm) in diameter in continuous lengths of 660 feet (201 m). The winch is supplied with 1320 feet (402 m) of cable. A traveler is shown in Figure 12.6.

Figure 12.6. Traveler in operation on a grass crop. (Photo courtesy Lockwood Corp.)

Other sizes and lengths of hose are available but are not commonly used. Sprinklers are operated through part of a circle covering 270 to 300° in such a manner that the lane ahead of the traveler remains relatively dry. Figure 12.7 is a schematic diagram of a typical traveler layout.

Traveler irrigation systems are typically designed to meet the crop and soil conditions outlined in Table 12.5.

The type and make of sprinkler are selected on the basis of cost and performance based on field tests of flow rate, operating pressure, wind speed, and lane spacing. Typical data are given in Table 12.6.

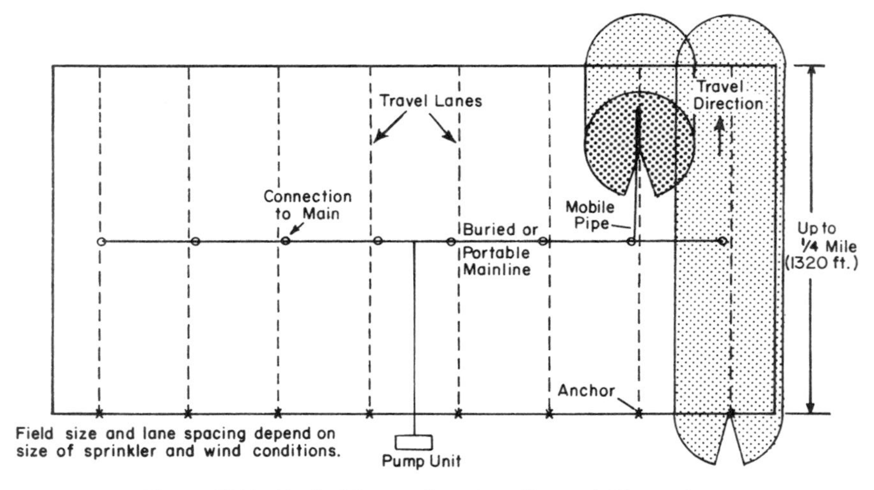

Figure 12.7. Typical layout for a traveling sprinkler system.

Table 12.5. Operating Data for Typical Traveler System

Area Covered, acres[a]	Soil Type	Irr. Cycle, Days	Flow Rate, gpm[b]	Water Applied, in.[c]
80	sandy loam	5	400	1.25
		5	500	1.50
		5	575	1.75
		5	660	2.00
80	clay loam	10	400	2.5
		10	500	3.0
		10	575	3.5
		10	660	4.0

[a]Ac = 0.405 ha
[b]gpm = 0.0631 l/sec
[c]in. = 2.54 cm

The information from Table 12.6 is integrated with the traveler winch performance data to give a system performance table useful in planning system operating modes. A typical performance table is shown in Table 12.7.

Table 12.6. Design Data for a Typical Traveler Sprinkler Nelson[a] Big Gun Model P2004 24° (courtesy Lockwood Corp.)

Discharge, gpm	Ring Size, inches	Pressure at gun, psi	Wetted Dia. 24° Angle, ft	Lane Spacing, ft		
				Winds Over 10 mph[b]	Winds 3-8 mph[c]	Winds Under 3 mph[d]
300	1.29 1¼	85	362	180-200	220-235	250-270
400	1.46 13/8	85	406	200-220	240-260	280-300
450	1.56 1½	85	417	210-230	250-270	290-310
500	1.56 1½	100	436	220-240	260-280	305-325
550	1.66 15/8	90	450	225-245	270-290	315-335
600	1.74 13/4	85	450	225-245	270-290	315-335
660	1.74 13/4	100	470	235-260	280-305	330-350

[a]L. R. Nelson Co., Inc., Brimfield, Ill. 61517.
[b]Wetted diameter is 50 to 55%.
[c]Wetted diameter is 60 to 65%.
[d]Wetted diameter is 70 to 75%.

Example

In an example system design calculation assume the following:

pumping rate = 550 gpm at 90 psi (34.7 l/sec at 6.3 kg/cm^2)
disposal site area = 80 acres (32.4 ha)
average application = 1.0 in. (2.54 cm)
wind speed = 2-4 mph (3.2-6.4 km/hr)

Find the lane spacing, travel speed, and irrigation time.

Note from Table 12.6, the sprinkler performance data suggests a sprinkler lane spacing of 290 to 315 ft (88.4 to 96.0 m). Using the data in Table 12.7, a lane spacing of 310 feet (94.5 m) is recommended, which gives the following operating data:

area sprayed = 9.4 ac/¼ mile travel (3.8 ha/0.4 km travel)
travel speed = 3.0 ft/min for a 1.0 in. average application (0.91 m/min for a 2.54 cm average application)
spraying rate = 1.26 ac/hr (0.51 ha/hr)

Table 12.7. Traveler Performance Data[a]

Discharge, gpm	Lane Spac., ft	Ring Size, inches	Area Irrig. per ¼ Mile, acres	Area Irrigated, ac/hr and Depth Applied, in.[b] Travel Speed, feet per minute							
				0.6	1.0	1.5	2.0	2.5	3.0	3.5	4.0
	250		7.6								
300	260	1-1/4	7.9	0.22 (3.1)	0.36 (1.8)	0.53 (1.2)	0.72 (0.9)	0.90 (0.7)	1.07 (0.6)	1.25 (0.5)	1.43 (0.45)
	270		8.2								
	260	1-3/8	7.9								
350	268		8.1	0.22 (3.5)	0.37 (2.1)	0.56 (1.4)	0.75 (1.1)	0.94 (0.8)	1.11 (0.7)	1.29 (0.6)	1.48 (0.5)
	278	1-1/4	8.4								
	270	1-1/2	8.2								
400	276	1-3/8	8.4	0.23 (3.9)	0.38 (2.3)	0.57 (1.6)	0.76 (1.2)	0.95 (0.9)	1.14 (0.8)	1.33 (0.7)	1.52 (0.6)
	286		8.7								
	280	1-1/2	8.5								
450	286		8.7	0.24 (4.2)	0.39 (2.5)	0.59 (1.7)	0.79 (1.3)	0.98 (1.0)	1.18 (0.8)	1.38 (0.7)	1.58 (0.6)
	294	1-3/8	8.9								
	290	1-3/4	8.8								
500	295	1-5/8	8.9	0.24 (4.5)	0.41 (2.7)	0.61 (1.8)	0.81 (1.4)	1.01 (1.1)	1.22 (0.9)	1.42 (0.8)	1.63 (0.7)
	302	1-1/2	9.2								
	300	1-3/4	9.1								
550	305		9.2	0.25 (4.8)	0.42 (2.9)	0.63 (1.9)	0.84 (1.5)	1.05 (1.2)	1.26 (1.0)	1.47 (0.8)	1.68 (0.7)
	310	1-5/8	9.4								
	310	1-7/8	9.4								
600	314	1-3/4	9.5	0.26 (5.1)	0.43 (3.0)	0.64 (2.0)	0.86 (1.5)	1.07 (1.2)	1.29 (1.0)	1.50 (0.9)	1.72 (0.8)
	318	1-5/8	9.6								
	318	1-7/8	9.6								
650	320	1-3/4	9.7	0.26 (5.4)	0.44 (3.3)	0.66 (2.2)	0.88 (1.6)	1.10 (1.3)	1.32 (1.1)	1.54 (0.9)	1.77 (0.8)
	322	1-5/8	9.8								

[a]Courtesy Lockwood Corp.

[b]Depth applied shown in parentheses.

From these data the following can be calculated:

$$\text{spraying time/lane} = \frac{(1320 \text{ ft})}{(3.0 \text{ ft/min})\ 60 \text{ min/hr}} = 7.3 \text{ hr}$$

$$\text{net}^{a} \text{ spraying time/80 ac} = \frac{80 \text{ ac}}{1.26 \text{ ac/hr}} = 63.5 \text{ hr}$$

$$\text{lanes required/80 ac} = \frac{80 \text{ ac}}{9.4 \text{ ac/¼ mile lane}} = 8.5$$

Travelers possess several unique capabilities that account for their success to date:

1. Because of the large orifice in the sprinkler [1-5/8 in. (4.13 cm) in the example] the system will pass large solids and has been used to spray municipal sludges and manures.
2. The winch is basically powerful enough to pull the system over any reasonable terrain that can be farmed. The lanes must, however, be straight and sharp brakes in contour may cause the unit to upset because of the straight cable pull.
3. Any size crop can be irrigated including tree crops.
4. Labor requirements and system costs are comparable to other portable systems.

The drawbacks to traveler systems, just as unique and perhaps severely limiting their use, particularly for spraying municipal wastewater and sludges, are:

1. Aerosol drift can be very extensive, with "mist" easily noted ½-mile (0.8 km) away during windy conditions.
2. Influence of the wind on pattern uniformity can be very significant.
3. Application rates are generally higher than any other system.
4. The hose is susceptible to damage and repairs are costly.
5. Lane maintenance problems are accentuated as heavy applications are applied to wet soil profiles (see Figure 12.8).
6. Spraying sludge and wastewater with this system creates onerous working conditions for the operating labor (see Figure 12.8).
7. Spraying sludges on cover crops may have a detrimental physiological effect.
8. Other operating concerns include the required use of an internal combustion engine to drive the system if the sprayed liquid is loaded with solid material.

---

[a] Gross spraying time would reflect the time required to move the system between lanes estimated at ½ to 1 hour per move.

Figure 12.8 shows a combination of factors that emphasize the limitations of a traveler. The high application rate has caused significant ponding on the field. Since the material being sprayed is sludge, it is probable that the soil intake rate has been reduced by soil plugging.

Figure 12.8. A traveler system spraying sludge on crop land.

CENTER PIVOT

Center pivot systems consist of a lateral carried on self-propelled towers that rotate about a permanent pivot point at which power and water are introduced. A typical system in irrigation service is shown in Figure 12.9.

The lateral pipe is commonly 6-, 6-5/8-, or 8-inch (15.2, 16.8, or 20.3 cm) diameter steel. Standard machines fitting into quarter sections are about 1300 ft (396 m) long and irrigate 130 acres (52.7 ha). Towers may be powered by oil (under hydraulic pressure to propel the tower), water or electricity. Electricity is preferred for spray irrigation of wastewaters. Electrically powered pivots are available with 8 to 10 towers per system requiring 126 ft (38.4 m) standard spans and up to 168 ft (51.2 m) stretch spans. Systems using 8 inch (20.3 cm) pipe and spray bars require 104 ft

Figure 12.9. A center pivot in irrigation service featuring under-truss design and powered by electric motors. (Photo courtesy of Lockwood Corp.)

(31.7 m) spans. The most common structural arrangement is the under-truss design with truss rods stabilized by "V" jacks.

Typical general specifications of the machines used on the Muskegon system are as follows:

Number of towers, 10
Four 104 ft by 8 inch pipe spans (31.7 m by 20.3 cm)
Five 124 ft by 6-5/8 inch pipe spans (37.8 m by 16.8 cm)
One 144 ft by 6-5/8 inch pipe span (43.9 m by 16.8 cm)
End boom length, 13 ft 5 inch (4.1 m)
System overall length, 1200 ft (366 m)
Flow rate, 1135 gpm (71.6 l/sec)
Pivot pressure, 44 psi (3.1 kg/cm^2)
Ground clearance, 9 ft (2.74 m)
Electrical power, 1 hp high-torque motors.

The ground speed, therefore time per revolution of electrically powered center pivots, is controlled by setting the running time on the end tower motor. Through the automatic alignment system, all intermediate towers will maintain an in-line position. The running time of the end motor is controlled by a percentage timer whose characteristics are shown in Table 12.8. This table is based on a wheel turning 0.67 rpm with 20.1 inch (51.1 cm) rolling radius (11 x 22.5 tires). The time required to make a full circle varies slightly with field conditions.

Table 12.8. Hours per Revolution[a]

Pipe Length, ft	Timer Setting Motor Run Time, %										
	100	90	80	70	60	50	40	30	20	10	5
668	9.2	10.1	11.5	13.1	15.3	18.4	23	30.5	46.0	92	184
792	11.0	12.2	13.8	15.8	18.4	22	27.5	37	55	110	220
916	12.8	14.2	16.1	18.2	21.5	25.5	32	43	64	128	256
1040	14.7	16.3	18.4	21	24.5	29.5	37	49	74	147	294
1164	16.6	18.5	21	24	27.5	33	41	55	83	166	332
1288	18.5	21	23	26	31	37	46	62	92.5	185	370
1412	20.2	23	25	29	34	41	51	68	100	202	404
1536	22	25	28	31	37	44	55	73	110	220	440

[a]Courtesy Lockwood Corporation.

From Table 12.8 a 1288-ft (393 m) pipe has a minimum time per revolution of 18.5 hours and maximum of 370 hours. To complete the calculation the data shown in Table 12.9 are required.

Example

An example of the use of Tables 12.8 and 12.9 follows.

Given: 10-tower system 1288 ft (393 m) long
Flow capacity, 1000 gpm (63.1 l/sec)
Time per revolution, 62 hours

Required: Application, in.
Timer setting, %

From Table 12.9, the application rate is 0.0168 in. per hour

Application = (62) (0.0168)
= 1.04 in./revolution (2.64 cm/revolution)

From Table 12.8, the required timer setting is 30%.

Given: 8-tower system 1040 ft (317 m) long
Flow capacity, 900 gpm (56.8 l/sec)
Application, 1.5 in. per revolution (3.81 cm per revolution)

Required: Time per revolution, hr
Timer setting, %

From Table 12.9, the application rate is 0.0227 in. per hr (0.058 cm per hr).

Timer per revolution = 1.5/0.0227 = 66 hours

From Table 12.8, the required timer setting is between 20 and 30%. By interpolation the correct value is 23.2%.

Table 12.9. Application Rate in Inches per Hour[a]

Number of towers	5	6	7	8	9	10	11	12
Length of pipe[b]	668	792	916	1040	1164	1288	1412	1536
Acres irrigated[c]	38.2	52.3	68.7	87.2	108	131	156.2	183.6
gpm	inches/hr							
400	0.0231	0.0169	0.0128	0.0101	0.0082	0.0067	0.0056	0.0048
500	0.0289	0.0211	0.0160	0.0126	0.0102	0.0084	0.0070	0.0060
600	0.0346	0.0253	0.0192	0.0151	0.0122	0.0101	0.0084	0.0072
650	0.0375	0.0274	0.0209	0.0164	0.0132	0.0109	0.0091	0.0078
700	0.0404	0.0295	0.0225	0.0177	0.0143	0.0118	0.0099	0.0084
750	0.0433	0.0310	0.0241	0.0189	0.0153	0.0126	0.0106	0.0090
800	0.0462	0.0337	0.0257	0.0202	0.0163	0.0135	0.0113	0.0096
850	0.0491	0.0358	0.0273	0.0217	0.0174	0.0143	0.0120	0.0102
900	0.0520	0.0379	0.0289	0.0227	0.0184	0.0151	0.0127	0.0108
950	0.0548	0.0400	0.0305	0.0240	0.0194	0.0160	0.0134	0.0114
1000	0.0577	0.0421	0.0321	0.0253	0.0204	0.0168	0.0141	0.0120
1050	0.0606	0.0442	0.0337	0.0265	0.0214	0.0177	0.0148	0.0126
1100	0.0635	0.0464	0.0353	0.0278	0.0225	0.0185	0.0155	0.0132
1150	0.0664	0.0467	0.0369	0.0290	0.0235	0.0193	0.0162	0.0138
1200	0.0693	0.0487	0.0385	0.0303	0.0245	0.0202	0.0169	0.0144
1250	0.0722	0.0508	0.0401	0.0316	0.0255	0.0210	0.0176	0.0150
1300	0.0750	0.0528	0.0417	0.0329	0.0265	0.0219	0.0183	0.0156
1400	0.0808	0.0590	0.0450	0.0354	0.0286	0.0236	0.0198	0.0168

[a]Courtesy Lockwood Corp.

[b]Dimensions of machine: 123 ft 8 in. (37.7 m) from pivot to first tower. 124 ft 4 in. (37.9 m) between all other towers. End boom is 40 ft long (12.2 m).

[c]These figures based on 100% application rates with the end gun operating continuously covering 60 ft (18.3 m) beyond the end of the system.

gpm = 0.0631 l/sec
ac = 0.405 ha
in. = 2.54 cm

The application rate allowable on a specific project depends on a range of considerations. If the application rate is too high the project could develop high runoff, which would damage the site and might pollute nearby streams. An initial effort has been made to provide guidelines based on the most important of these considerations, that is the soil type (see Table 12.10).

Table 12.10. Soil Classification and Maximum Application Rates for Center Pivots[a]

Soil Classification	System Capacity		Max. Application Rate, in./hr
	gpm[b]	gpm/acre	
Sand	1250	9.8	0.022
Loamy Sand	1100	8.6	0.019
Loam	960	7.5	0.017
Sandy Clay Loam	900	7.0	0.015
Silty Clay Loam	690	5.4	0.012
Clay	600	4.7	0.010

[a]Based on a sprayed area of 128 acres (51.8 ha).
[b]Source: SIA[10]

gpm = 0.0631 l/sec
ac = 0.405 ha
in. = 2.54 cm

Other project variables that affect the decision on design application rate are: (1) whether the system is equipped with spray bars having a high instantaneous application rate or impact sprinklers with lower rates, (2) general field slope as it relates to runoff hazards, (3) type of cover crop (sod crops tend to hold excessive moisture whereas row crops are more susceptible to runoff), and (4) project objective with disposal sites (as opposed to utilization sites) more susceptible to runoff hazards because of a more saturated soil profile.

As a further example of how these data may be used, consider the following system design:

Given: Soil, silty clay loam
Site area, 130 acres (52.7 ha)
Required application, 1.0 in. per revolution (2.54 cm per revolution)

Required: System capacity, gpm
Time per revolution, hrs
Timer setting, %

From Table 12.10, the maximum application rate allowable is 0.012 in./hr (0.030 cm per hr). From Table 12.9, a 10-tower system with a capacity of 700 gpm (44.2 l/sec) and an application rate of 0.0118 in./hr (0.030 cm/hr) is adequate.

$$\text{Time per revolution} = 1.0/0.0118 = 85 \text{ hours}$$

From Table 12.8 by interpolation, the timer setting is 22.5%.

A special consideration in the spraying of wastewater is the minimization of spray drift (aerosols) from the disposal site. For this reason, the center pivots installed at Muskegon were equipped with spray bars mounted as shown in Figure 12.10. With spray bars in this position it is possible to spray the effluent directly onto the cover crop. A close-up of the piping arrangement is shown in Figure 12.11.

Figure 12.10. Center pivot as installed at Muskegon. Note spray bars. (Courtesy Lockwood Corp.)

Pressure in the main pipe at the pivot was typically 30-50 psi (2.1-3.5 kg/cm^2), which was adequate to overcome friction losses and provide the required spray bar pressures. A hose connection was supplied between the main pipeline and the spray bar to allow for individual pressure settings. A floodjet type nozzle was used.

A significant drawback of spray bars is their high instantaneous application rate. The floodjet nozzles apply water in a curtain as they move across the field. Alternate nozzles are directed forward and backward in an effort to widen the application pattern. Typically, however, with an application of 1/2 in. (1.27 cm) per pass occurring in the 5 to 10 minutes that the pivot

Figure 12.11. Details of the spray bar attachment (Courtesy Lockwood Corp.).

sprays over a specific location, average instantaneous application rates of 3 to 6 in. (7.6 to 15.2 cm) per hour can be expected. It is possible through the use of standard irrigation sprinklers to widen the application pattern and reduce the average instantaneous application rate to perhaps 1/2 to 1-1/2 inches (1.3 to 3.8 cm) per hour. This greatly aggravates the aerosol problem, however.

There is an urgent need for the development of better spray nozzles and sprinklers. Sprinklers were developed to perform best in solid set systems mounted close to the ground. The conditions of spraying from a pivot are quite different and require a new type of sprinkler. Sprinklers should be designed for:

1. low angle (0-5°) because height is already available in the mounting location
2. low pressure to minimize aerosol drift and pumping energy. There is no need to throw the water to get coverage as outlets are available at about 10 ft spacing on the main pivot pipeline
3. the use of main drive nozzles only with no fill-in nozzles. Adequate pattern overlap exists with normal spacing.

Since a center pivot system is carried on wheels across the disposal site under wet conditions, attention must be directed at the ability of the system to avoid deep rutting, which eventually mires the wheels. Experience has shown that if pneumatic tires are used and ground pressures are sufficiently low, a satisfactory operation can be expected. Maximum allowable contact pressures related to soil type are shown in Table 2.11.

Table 2.11. Maximum Allowable Ground Contact Pressure as a Function of Soil Type.

Fines,[a] %	Max. Allowable Ground Contact Pressure, psi
20	25
40	16
50	12

[a]Defined as material passing a 200 mesh sieve.
psi = 0.07 kg/cm^2

Practically, Table 12.11 translates into a contact pressure of about 12 psi (0.84 kg/cm^2) for clay soils up to 20 to 30 psi (1.41 to 2.11 kg/cm^2) for sandy soils. The design ground contact pressure also relates to root zone moisture content at time of spraying. If effluents or sludges must be applied on a wet soil profile, the allowable ground pressure must be reduced. However, it may be possible to use spray bars applying effluent behind the machine. This keeps at least one tower wheel operating on relatively dry ground. A second possibility is the use of drop hoses placing the effluent directly on the ground surface, with the area near the tires purposely left dry. A combination of tire size and number of towers must be selected to give a satisfactory ground pressure when carrying a spraying machine.

Terrain can limit the suitability of a site primarily because of the runoff hazard. Mechanically, the electric drive pivots have the power and flexibility to handle rolling terrain with slopes of 15 to 20%.

An assortment of structural considerations are of a secondary concern to the system designer. They include:

1. Pivot and pivot anchor. A suitably sized and reinforced concrete foundation should be installed. The pivot should be designed to AISC specifications and certified by a registered professional engineer.
2. System loading. The design wind load should be as generated by a 90 mph (145 km/hr) wind with a 1.1 safety factor (SIA[10]). The structure should be designed to AISC specifications by a registered professional engineer and capable of withstanding the wind loads with no water in the pipe.
3. The system, particularly the main pipeline, should be suitably protected from the corrosive effects of the effluent. All manufacturers provide some type of protective coating varying from galvanizing, epoxy coating and painting to low-alloy steels.
4. Since drive train components are susceptible to wear, a suitable warranty is available from most manufacturers. The SIA[10] guide recommends a 600-hour warranty commencing with acceptance of the system.

PUMPING AND PIPELINE CONSIDERATIONS

The design and selection of a pump for spraying wastewaters involves considerations similar to pumping irrigation waters with one notable exception: the requirement to pump sludges or chopped solids. The pump selected must obviously meet such system requirements as flow rate and pressure. In addition for all systems using small orifice sprinklers or nozzles, the maximum particle size allowable must be determined and adequate screening or filtering provided to eliminate plugging. If this is accomplished, conventional high-pressure centrifugal irrigation pumps can be used. There are many good references on the design of this type of pump.[20] An example of such an installation is shown in Figure 12.12.

The installation in Figure 12.12 features an electrically powered centrifugal pump with a hand priming pump and a check valve and gate valve on the discharge side. If the unit were mounted in a "wet" sump below the lagoon water surface, it could be fitted with automatic electrical controls and be self-priming.

If some solids loading is anticipated, the closed impeller of the high-pressure pump must be modified to an open or semi-open impeller with a vane clearance adequate to pass the maximum anticipated particle size. When the bolt-on inspection cover is removed, the impeller is fully exposed, allowing for removal of clogged materials. In addition, pumps handling solids should be provided with easily removable cover plates. Figure 12.13 shows such a pump handling sanitary sludge that is being sprayed through travelers.

It is possible to pump liquid manures and sludges with solids contents of 8-10% by weight.[21] Since this requires the use of open impellers with

Figure 12.12. High-pressure irrigation pump installation. (Courtesy Gorman Rupp Co., Mansfield, Ohio 44903).

Figure 12.13. Diesel-powered centrifugal pump supplying sludge to a traveler. (Courtesy Gorman Rupp Co.)

relatively low head per stage capability, multistage pumping units are used. Two 4-in. (10.2 cm) solids handling pumps in series are shown in Figure 12.14. Note also the cleanout ports on both pumps to allow maintenance work on the impeller without disturbing the pipeline hookup.

Figure 12.14. Two 4-inch solid handling pumps in series. (Courtesy Gorman Rupp Co.).

A relatively new development is the chopper pump. This unit uses a submerged open-end suction impeller centrifugal pump preceded by a rotating cutter assembly. The cutter assembly consists of a rotating knife cutting against a stationary plate. This assembly is mounted directly to the inlet side of the impeller.

These pumps have successfully handled flows from a wide assortment of industrial and sanitary wastes. They are easy to install, have a good life and low maintenance cost. They are occasionally used as circulating pumps to keep lagoon and sump solids agitated and in suspension.

Pipeline materials generally available include plastic, asbestos cement, steel, and aluminum.

Plastic pipe is usually polyvinylchloride (PVC) or polyethylene (PE). It is used widely in irrigation systems because it is light in weight, easy to handle, relatively inexpensive, and resists deterioration well. But its disadvantages include low mechanical strength and some deterioration from sunlight. It is not suited to portable systems and must be buried, carefully backfilled, and supported by reaction blocks where necessary. Coupling methods include solvent-welded fittings, ring-gasket bell- and spigot joints, and chevron gaskets in epoxy-coated steel fittings.

Asbestos-cement pipe has excellent resistance to attack from most wastewater chemicals. It is well-suited to buried permanent installations where long life is important. Installation is somewhat difficult because of the weight of the pipe. Bell and gasket couplers require reaction blocks at all tees, elbows, and end plugs. Backfilling must be done carefully to avoid rolling large rocks onto the pipe and cracking it. Asbestos-cement pipe is especially susceptible to damage caused by pressure surges. Special care must be taken to provide air bleed and vacuum relief valves and to use slow reacting control valves. In addition, care must be taken in starting and stopping pumps to avoid excess pressure build up. In extreme cases it may be necessary to start the pump against a slow reacting control valve in the closed position. The valve can then be opened slowly to the operating pressure.

Steel pipe is well-suited to a variety of applications. It may be installed in buried or above-ground systems. Coupling methods include welds, threads, flanges, or "Victaulic" couplers. Steel pipe is only as good as its coating, however, and the coating must be selected to resist the deteriorating influence of the anticipated effluent. Asphalt-dipped and felt-wrapped pipe is commonly used in irrigation systems. "Victaulic" couplings are more expensive but better suited to above-ground installation where pipelines may have to be dismantled and moved.

Aluminum pipelines are lightweight, easy to handle and well-suited for above-ground portable systems. Clad aluminum pipe will resist deterioration for most irrigation service. However, some wastewater chemicals can seriously limit the useful life of aluminum irrigation piping. This should be checked by running deterioration tests on the actual wastewaters to be sprayed on a given project. Some attempts have been made to coat aluminum pipe for buried applications with limited success. Few aluminum pipelines have been buried, however. A wide assortment of quick couplers and valves are available for aluminum piping systems.

Additional components to be included in the piping network include:

1. Shut-Off Valves. These are usually butterfly or gate valves used to isolate sections of the piping network as appropriate to meet sprinkler schedules or to isolate filters or control valves for maintenance work without shutting the system down. Gate valves are not normally designed for throttling service.

2. Pressure or Flow Control Valves. These are usually globe valves with a diaphragm-actuated disc. The diaphragm reacts to a pilot circuit sensing pressure or flow. These valves are used to maintain correct sprinkler operating pressures. They may also be actuated by emergency or routine controlled shut-down signals.
3. Air Vent and Vacuum Relief Valves. These valves are essential to all systems. The air vent valves are located at all system high points. They vent trapped air, thus avoiding air locks and surging caused by air escaping through valves and sprinklers. The vacuum relief valve relieves negative pressure in the system. This prevents the collapse of large-diameter pipe and prevents pulling soil back into gasketed joints and couplings.
4. Screening and Filtering. Devices should be included wherever needed to prevent plugging of other system components.

SYSTEM OPERATIONAL CONSIDERATIONS

Personnel involved in a spray disposal system operation are usually drawn from backgrounds other than agricultural. They work for organizations concerned with waste disposal problems, but usually they lack the sensitive judgment required to operate farm disposal sites properly. Their organizations, concerned with other matters, are likely to consider waste disposal as a "necessary evil" forced upon them by government regulations. In this context, then, there is the extreme likelihood that inadequate attention will be paid to the details of properly operating the irrigation system and the farming and cropping system.

In small communities, the individual operating the disposal site may also be the superintendent of public works and, therefore, responsible for such things as operation of the water supply system, garbage collection, and street maintenance. This argues strongly for a well-planned, fully automatic system. Agricultural procedures should be developed with input from local farmers.

SYSTEM COSTS

Caution must be exercised when developing system costs to be sure they reflect conditions on site. At present, the design of wastewater systems reflects two differing standards. One standard involves the use of agricultural design philosophies and the other involves civil engineering standards. In the area of costs, the standards are significantly in conflict. As an example, solid-set systems on agricultural crops typically cost about $500/acre. When installed to civil engineering standards on wastewater systems, however, costs as high as $4000 to $5000/acre have been reported.[19] This apparently reflects differing equipment specifications such as the use of cast iron pipe instead of the more economical PVC plastic pipe. The same contrast is reflected in operating labor with variations of 5:1 in hourly rates. The system designer must know local conditions intimately if his capital and operating costs are to be realistic.

REFERENCES

1. Hunt, P. "Microbiological Responses to the Land Disposal of Secondary Treated Municipal-Industrial Wastewater," Special Report No. 171, *Wastewater Management by Disposal on the Land* (Hanover, N.H.: Corps of Engineers, 1972), Chapter 5.
2. Murrmann, R. P. and F. R. Koutz. "Role of Soil Chemical Processes in Reclamation of Wastewater Applied to Land," Special Report No. 171, *Wastewater Management by Disposal on the Land* (Hanover, N.H.: Corps of Engineers, 1972), Chapter 4.
3. Peterson, M., *et al.* "A Guide to Planning and Designing Effluent Irrigation Disposal Systems in Missouri," U. of Missouri Extension Division, Publ. MP 337 3/73/1250, Columbia, Mo. (1973), p. 16.
4. Pennsylvania Department of Environmental Resources. *Spray Irrigation Manual*, Publication No. 31 (Harrisburg, Pa.: Bureau of Water Quality Management, 1972).
5. Sopper, W. E. "Crop Selection and Management Alternatives–Perennials," *Proceedings Recycling Municipal Sludges and Effluents on Land* (Washington, D.C.: National Associations of State Universities and Land Grant Colleges, 1973), pp. 143-153.
6. Rickard, W. "Botanical Components Involved in Land Disposal of Wastewater Effluent," Special Report No. 171, *Wastewater Management by Disposal on the Land* (Hanover, N.H.: Corps of Engineers, 1972), Chapter 6.
7. Norum, E. M. "Development of Land Disposal Equipment Special System Design Considerations," *ASAE Proceedings North Carolina Irrigation Conference*, Raleigh, N.C., November 28 and 29, 1973. p. 93.
8. Christiansen, J. E. "Irrigation by Sprinkling," Bulletin 670, U. of California College of Agriculture, Agr. Exp. Station, Berkeley, California (1942).
9. Soil Conservation Service. "Sprinkler Irrigation," *National Engineering Handbook,* Section 15, Chapter 11 (Washington, D.C.: Soil Conservation Service, 1960).
10. Sprinkler Irrigation Association. "Guide to the Use of Center Pivot Irrigation Equipment for Waste Treatment of Effluent," Silver Springs, Maryland (1975).
11. Norum, E. M. "A Method of Evaluating the Adequacy and Efficiency of Overhead Irrigation Systems," ASAE Paper No. 61-206 presented at Annual Meeting, Ames, Iowa (1961).
12. Norum, E. M. "A Suggested Procedure for Characterizing the Performance of Spray Disposal Equipment," ASAE Paper No. 74-2553 presented at Annual Meeting, Chicago, Illinois, 1974.

13. American Society of Agricultural Engineers. Standard S263.1, St. Joseph, Michigan (1962).
14. Myers, E. A. "Sprinkler Irrigation Systems–Design and Operation Criteria," *Symposium Proceedings on Recycling Treated Municipal Wastewater and Sludge Through Forest and Cropland.* (University Park, Pa.: The Pennsylvania State University Press, 1973), p. 324.
15. Frost, T. P., R. E. Towne, and H. J. Turner. "Spray Irrigation Project, M. T. Sunapee State Park, New Hampshire. *Symposium Proceedings on Recycling Treated Municipal Wastewater and Sludge Through Forest and Cropland.* (University Park, Pa.: The Pennsylvania State University Press, 1973), p. 385.
16. Williams, T. C. "Utilization of Spray Irrigation Study for Wastewater Disposal in Small Residential Development," *Symposium Proceedings on Recycling Treated Municipal Wastewater and Sludge Through Forest and Cropland.* (University Park, Pa.: The Pennsylvania State University Press, 1973), p. 385.
17. Environmental Protection Agency. "Engineering Feasibility Demonstration Study for Muskegon County, Michigan, Wastewater Treatment–Irrigation System," Report No. 11010 FMY 10/70 U.S. Government Printing Office (1973).
18. Bauer, W. J. and D. E. Matsche, "Large Wastewater Irrigation Systems: Muskegon County, Michigan and Chicago Metropolitan Region," *Symposium Proceedings on Recycling Treated Municipal Wastewater and Sludge Through Forest and Cropland.* (University Park, Pa.: The Pennsylvania State University Press, 1973), p. 345.
19. Williams, T. C. "Recycling Municipal Sludges and Effluents on Land," *Proceedings Recycling Municipal Sludges and Effluents on Land* (Washington, D.C.: National Associations of State Universities and Land Grant Colleges, 1973), p. 169.
20. Woodward, G. O. *Sprinkler Irrigation.* (Silver Springs, Md.: Sprinkler Irrigation Assoc., 1959).
21. Bohley, P. B. "Pumps and Equipment for Sanitary and Agricultural Wastes," *Proceedings North Carolina Irrigation Conference*, Raleigh, N.C., November 28 and 29, 1973.

13

REVIEW OF MUSKEGON COUNTY WASTEWATER MANAGEMENT SYSTEM

Edward M. Norum

Chief Product Engineer
Lockwood Corporation
Gering, Nebraska 69341

INTRODUCTION

This review of the results to date (April, 1975) of the Muskegon County Wastewater Management system must be considered preliminary because the project has yet to reach full design capability particularly with respect to irrigation and cropping. Some of the project objectives and concerns will take years to appraise accurately. Emphasis is placed, herein, on the irrigation equipment, which is an essential component in the system.

OVERALL PROJECT OBJECTIVES

Much of the basic motivation for the Muskegon system was derived from the findings of the Lake Michigan Enforcement Conference.[1] Concern for the pollution of the waters of Lake Michigan led to several recommendations for the Muskegon system reported by EPA[1] in the final draft of the Environmental Impact Statement:

> By December 1972, the respective states shall require municipalities and industries to achieve at least 80% reduction of total phosphorus and to produce an effluent that will not result in degradation of Lake Michigan's water quality.
>
> Unified collection system service contiguous urban areas will be encouraged.

> Discharge of treatable industrial wastes (following any needed preliminary treatment) to municipal sewer systems will be encouraged.
>
> Programs providing maximum use of area-wide sewage treatment facilities (to discourage the proliferation of small treatment plants in contiguous urbanized areas) will be encouraged.
>
> The replacement of septic tanks with adequate collection and treatment facilities will be encouraged.

It is appropriate to judge the relative success of the project by the degree to which the objectives are being met. Although the project in large measure preceded the Federal Water Pollution Control Act Amendments of 1972, it will inevitably be compared to the objectives of the act. In particular, Section 201 of the act declares:

> Waste treatment management plans and practices shall provide for the application of the best practicable waste treatment technology before any discharge into receiving waters, including reclaiming and recycling of water, and confined disposal of pollutants so they will not migrate to cause water or other environmental pollution and shall provide for consideration of advanced waste treatment techniques.

More recent EPA directives[2] relate to land application of wastewaters as follows:

> All feasible alternative waste management systems shall be initially identified. These alternatives should include systems discharging to receiving waters, systems using land or surface disposal techniques, and systems employing the reuse of wastewater.

It is then against these growing imperatives that the Muskegon County Board decided to develop the Muskegon County wastewater plan. In doing so they avoided the more traditional solutions in favor of transferring the treatment facility away from Lake Michigan and utilizing the wastewater and the land for growing crops. It was a courageous and farsighted decision. Much is at stake in the success of the project.

In the formulation of the Muskegon County Wastewater Management (MCWM) plan, three interrelated environmental principles were considered:[1]

1. The environment is a single system with land, air, water and living resources interacting and affecting and being affected by human activities.
2. The environmental system is closed; nothing can be "thrown away." No more "dumps" exist. The concept of complete recycling must be developed.
3. Some pollutants are potential resources out of place. If relocated in the environmental system, they may take on a new value.

After completion of the project a wide variety of evaluation studies will be conducted for periods of several years. Performance of the project will be measured in several ways:[1]

1. water quality monitoring of surface and groundwaters affected by the project
2. evaluation of performance of treatment components
3. evaluation of agricultural productivity, soils effects, and economic benefits of wastewater irrigation
4. evaluation of social, economic, and environmental impact of project on the community.

PROJECT PHYSICAL DESCRIPTION AND COSTS

The scientific background and construction features of the project have been well documented (for example, Muskegon County Board,[3] Bauer,[4] and Wilson[5]). In addition, there has been generally favorable coverage in periodicals (for example, Chaiken, *et al.*[6] and Forestell[7]). The somewhat unique problems of acquiring the project site were described by Postlewait.[8] The Muskegon system, as shown schematically in Figure 13.1, consists of the following major components: (1) intercepting sewage facilities, (2) lagoon treatment and storage facilities, (3) irrigation facilities,

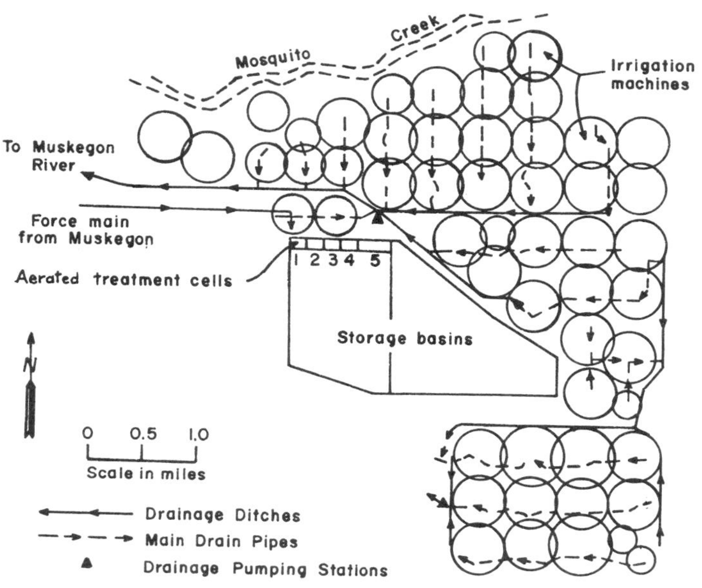

Figure 13.1. Wastewater flow at Muskegon (after Chaiken, *et al.*[6]).

and (4) drainage facilities. The total capacity of 42.0 mgd (393,120 m^3/day ha) is based on 1992 requirements. The interception sewerage facilities consist of seven pumping stations, the largest having a capacity of 88 mgd (823,680 m^3/day ha) and pumping the untreated wastewaters through 11 miles (17.7 km) of 66 in. (168 cm) force main to the treatment site. The treatment site, 17 miles (27.4 km) east of Lake Michigan, consists of 10,800 acres (4374 ha) in lagoon and irrigation facilities. Pretreatment begins with aeration in three 8-acre (3.24 ha) lagoons with a detention time of three days. Following aeration, the effluent is conveyed to either of two 850-acre (344 ha) storage lagoons, which have a minimum sludge storage depth of two feet (0.61 m) and an initial working depth of nine feet (2.74 m). The initial storage volume is about 5,100 mg (19,300,000 m^3) which provides 150 days of storage of current flows to permit water to be sprayed only during the growing season. Before spraying, the wastewaters are chlorinated in two channels that provide a minimum of 15 minutes contact time. Dosage rates up to 14.4 mg/l are possible.

Two multi-unit irrigation pumping plants supply water through irrigation pipelines to the 55 irrigation machines. The center pivot irrigation machines (also described in Chapter 12) have the following general characteristics:

Overall length, 700-1400 ft (213-426 m)
Flow rate, 380 to 1530 gpm (24.0 to 96.5 l/sec)
Operating pressure, 35 to 84 psi (2.46 to 5.90 kg/cm^2)
Application rate, 0.0239 in./hr (0.061 cm/hr) or 4.0 in./wk (10.2 cm/wk) (continuous operation)

The drainage facilities consist of a network of 6- or 8-in. (15.2 and 20.3 cm) perforated plastic drain tiles augmented by drainage wells. The drainage water can be directed to two creeks on the site. A number of observation wells were drilled on the site to monitor groundwater quality and to study seepage from the large storage lagoons. A summary of the system construction costs is given in Table 13.1. Based on an ultimate capacity of 42 mgd (159,000 m^3/day), the unit cost for the Muskegon-Mona Lake land treatment facility is $626,000 per mgd ($165 per m^3/day).

Table 13.2 shows Cowlishaw's[9] cost comparison between the Muskegon County system and the Advanced Waste Treatment (AWT) facility at South Lake Tahoe. The economics reflected in the low operation and maintenance (O and M) costs are of special interest to the users as they must bear this cost directly for the life of the project. The rates are so nominal that the County Board could decide on a higher rate and use the surplus funds to help retire the capital obligations.

Table 13.1. Construction Contracts and Costs for the Muskegon System[a]

Contract No.	Description	Construction Costs, $
1	Clearance and site improvements	$1,915,000
2	6- and 8-in. (15.2 and 20.3 cm) drain lines	758,000
3	Drain lines over 10 in. (25.4 cm)	1,878,000[b]
4	Roadway culverts	108,000
5	Ditches and channels	1,178,000[b]
6	Irrigation pipelines	1,444,000
7	Admin. & chlorination facilities[c]	369,000[b]
8	Electrical power distribution	556,000
9	Observation and drainage wells	386,000
10	66 in. (167.6 cm) force main	4,684,000[b]
11	Force mains and sewers	1,220,000
12	Force mains and sewers	845,000
13	Force mains and sewers	1,132,000
15	Pumping station "C"	1,746,000[b]
16	Other pumping stations	1,312,000
17	Irrigation pumping stations	437,000[b]
18	Lagoons and treatment facilities	9,568,000[b]
22	Irrigation machines	1,353,000[b]
24	Utility relocations	298,000
	Subtotal	$31,187,000
Adjustments		
Plus:	Additional construction and storage	1,175,000
	Land costs	4,900,000
	Relocation costs	1,160,000
	Subtotal	$38,422,000
Less:	Intercepting sewerage facilities	11,004,000
	Whitehall-Montague treatment costs	1,139,000
Muskegon-Mona Lake Treatment Total		$26,288,000[d]

[a]After Cowlishaw[9]

[b]Subject to final adjustment

[c]Includes lab equipment and furniture

[d]Does not include financial, legal, or engineering costs

Table 13.2. Cost Comparison, Muskegon County and South Lake Tahoe (AWT)

Cost	Annual Cost[b] for 42 mgd (159,000 m^3/day) Plant, ¢/1000 Gallons (3.79 m^3)[a]	
	Muskegon County[c]	South Lake Tahoe[d]
Capital	23.2	24
Operation and Maintenance	7.3[e]	25
Total	30.5	49

[a]After Cowlishaw[9]
[b]Adjusted to October, 1974 (ENR CCI 2350)
[c]15% added for engineering, legal, financial, and administration costs. Capital costs reflect 20-yr period at 7%.
[d]Adjusted to comparable cost index and economics of scale factor.
[e]Does not reflect total anticipated income from farm crops, which could potentially offset most of the operation and maintenance costs.

RESULTS TO DATE

Center Pivot Wastewater Irrigation Machines

The prime component of the land treatment system is the wastewater irrigation machine. For this reason, the management firm contracting for the installation and start-up of the system (Teledyne Triple R Corp., Muskegon, Mich.) developed an Irrigation Equipment Optimization Program to study machine design parameters in detail. This program led to the installation and full-scale field tests of two center pivots. As reported by Hall,[10] the machines were evaluated for the following factors:

1. structural design
2. mechanical operation
   a. stability
   b. operating characteristics
   c. reliability
   d. wheel ruts
   e. anticollision device
3. water distribution
   a. nozzle selection
   b. uniform distribution of water
   c. maximum water application rate
   d. aerosol drift

Because of the comparatively high flow rate and low operating pressure, 8-in. OD (20.3 cm) main pipe was required for most of the length with some 6-5/8-in. OD (16.8 cm) pipe used on spans toward the end. The weight of the 8-in. OD (20.3 cm) pipe limited the spans to 104 ft (31.7 m).

The 6-5/8-in. OD (16.8 cm) pipe spans were 124 and 144 ft (37.9 and 43.9 m). Detailed structural analyses were carried out to confirm design strength. This work followed AISC procedures, was verified by a professional engineer, and is currently suggested by SIA.[11]

Mechanical stability was studied by operating the machines over dirt mounds at simulated slopes of 30%. This resulted in specifications for an extra-wide wheel base of 13 ft (3.96 m), which, according to Hall,[10] provides for overturning stability in a 70 mph (113 km per hr) wind with a 1.5 safety factor.

The following operating characteristics were studied to insure compatability with system requirements: (1) ease of starting and stopping the machine, (2) frequency of maintenance, (3) ease of making major repairs in the field, and (4) operational limitations imposed by weather. The Lockwood Corporation unit eventually specified has a monitoring control center including the full range of control and trouble lights normally found on irrigation pivots. Frequency and ease of making repairs can be established only after an extended period of operation. Hall[10] reported that the Lockwood Corporation unit became inoperable at 32°F (0°C) in a 49 mph (79 km per hr) wind because of ice accumulation. The manufacturer recommends an automatic low temperature shutdown set for 38°F (3.3°C) to avoid possible structural damage to the machine.

Because the units will make a large number of revolutions per year in relatively wet conditions, a possible rutting problem was investigated. A comparison of steel versus rubber tires was included. Steel wheels eliminate the puncture problems, but because of their relative inflexibility, mud continues to accumulate on the tread instead of chipping off. Hall[10] reported rut depths ranging from 4 to 16 in. (9.2 to 40.6 cm) after a full season of irrigation service with no apparent difference between the steel wheels [42 in. (106.7 cm) OD by 10 in. (25.4 cm) wide] and the pneumatic tires [11.2 by 24 in. (28.4 by 61.0 cm), 4 ply]. However, the machines with pneumatic tires [14.9 by 24 in. (37.8 by 61.0 cm), 6 ply] were eventually specified. Rutting remains a concern in selected areas of black loamy sand and will have to be studied continuously.

As shown in Figure 13.1, some of the machine application circles overlap. To prevent mechanical damage to the units should they overtake each other, an anticollision device was developed by Lockwood Corporation.[a] As a trailing machine encounters another machine it shuts down and waits four minutes. It then automatically resumes operation. Hall[10] reported satisfactory field tests, and the anticollision device was provided on machines as needed.

---

[a]Lockwood Corporation, Gering, Nebraska 69341

The nozzles selected for all machines were of a "flood jet" type. The nozzle sprays water over an operating range of 3 to 60 psi (0.21 to 4.22 kg/cm^2), and a full range of sizes from 0.016 in. (0.04 cm) diameter to 0.703 in. (1.79 cm) diameter are available. The nozzles were mounted to spray the water in a fan pattern behind the machine. Jet pattern overlap was provided to give a full curtain of application over the length of the machine. The nozzles have no internal vanes on which foreign material could accumulate. The basic fan distribution pattern develops over an external baffling surface that tends to remain clean because of the scouring action of the jet. The weakness of the nozzle lies in two features:

1. The water distribution across the fan is nonuniform, necessitating a design with excessive overlap to insure reasonable overall uniformities.
2. The water is deposited in a "sheet" manner, resulting in very high instantaneous application rates. There is a need for a nozzle that deposits water uniformly over a wide rectangular area with a minimum of aerosol drift. This would reduce the instantaneous application rate.

The flood jet nozzle was selected also because it operates well at low pressures giving a fairly coarse drop spectrum which resists formation of aerosols.

An effort was made in the development phase to provide a Christiansen's Uniformity Application Coefficient[12] of 85%. Hall[10] described a method for calculating the placement of cans in a radial pattern and to evaluate Christiansen's[12] coefficient directly. This method of can location fails to reflect the difference in area represented by each individual can. Hall's[10] results showed the Lockwood unit to have a uniformity of about 81%. Subsequently, SIA[11] guidelines developed an extension of Hall's method to reflect a procedure for weighting the can catchments to reflect area differences. The guidelines suggest a minimum uniformity coefficient of 80% is achievable by all pivot manufacturers. Both approaches are satisfactory as equipment development tools, but as acceptance tests they leave much to be desired. They are essentially one-dimensional tests and fail to reflect the influence of wind on the application pattern, the influence of topography on the application pattern, and the influence of machine operating characteristics on the application pattern as, for example, the tendency for pivots to slow down somewhat when moving up hill or into high winds.

The maximum allowable application rate measures the specific ability of the soils and crops selected on the disposal site to tolerate the mechanical abuse from the drops. Since soils are unique, planning for any large-scale project should include sprinkler infiltrometer tests on the disposal site and a determination of allowable application rates. Hall[10] found the average infiltration rate to be 10 to 14 in./hr for Muskegon soils. Hall[10] reported that this was less than the instantaneous rate of the machine and that it resulted in minor runoff and ponding. If serious problems developed during operation, he recommended throttling the flows back to reduce the instantaneous application rate.

Tests were made to determine the effect of pivot applications on small corn seedlings. Hall[10] reported no apparent damage to seedlings 4 inches high over a range of timer settings from 5-100%.

Aerosol drift determinations were made on the site and reported by Hall.[10] He recorded aerosols of 242 micron size, 1500 ft (457 m) downwind of the pivot (actual pivot point of rotation) at a height of 15 ft (4.6 m) in winds of 22 mph (35.4 km per hr). He recommended lowering the spray bar from 7 ft (2.1 m) as a means of minimizing the drift. He also noted that aerosol drift is related to a complex series of variables (such as terrain, temperature, humidity, thermal uplift, and wind turbulence) unique to each site. Planning for a large-scale system should include an on-site study of drift problems with buffer zones and wind breaks designed accordingly.

On the project operation, Cowlishaw[9] reported that the machines were phased into operation in the summer of 1974 until 54 units were operating by August. Because of a need to draw the reservoir down rapidly, irrigation rates were 4 in. (10.2 cm) per week, reflecting continuous machine operation even during rainy weather.

The problems encountered included: (a) a plugging of nozzles from weeds growing in the pump supply ditches, (2) some rutting problems on selected soil areas, and (3) premature deterioration of the electrical cable supplying the center pivots. High flotation tires will be tried to overcome the problem in Item 2. Special electrical cables may be required by the strenuous duty and continuously wet soils to avoid the problem in Item 3.

General Project Results

General results are limited because of the early stage of the project. Comments, however, are included on odor problems, lagoon operation, cropping results, drainage water quality, and irrigation pipelines.

Probably the most sensitive public relations concern associated with land treatment systems is the odor problem. There have been innumerable public hearings and lawsuits concerning odors from various projects, such as the Fulton County Sludge Disposal Project. At Muskegon, a high percentage of the wastewater flow is industrial, some of it from a paper mill. As reported by Cowlishaw,[9] the odors derive from the mercaptans and reduced-sulfur compounds present in the wastewater. Threshold odor concentrations for these compounds are measured in parts per billion, so they are readily detectable. A lawsuit has been filed against the county in an effort to force resolution of the odor problem. Cowlishaw[9] reported that the county responded in September, 1974, by adopting an order forcing the paper mill to stop discharging "any malodorous substances capable of creating a public nuisance." This order also served notice on wastewater dischargers of the county's intention to levy a surcharge to cover incremental treatment costs on constituent levels above the levels given in the service agreement. The aeration lagoons do not appear to create the problem, but, on the contrary, help to prevent odor problems in the storage lagoons. Performance of the aerated lagoons is reported by Cowlishaw[9] and is shown in Table 13.3.

Table 13.3. Aerated Lagoons, Average Performance Values

Parameter or Constituent	Influent	Aerated Lagoon Effluent	Per Cent Removal
BOD_5, mg/l	210	60	71
COD, mg/l	560	320	43
SS, mg/l	270	140	48
pH	7.5	7.5	–
Ammonia nitrogen, mg/l	5.5	2.0	64
Nitrate and nitrite nitrogen mg/l	0	0.2	–
Total Kjeldahl nitrogen, mg/l	9	8	11
Orthophosphate, mg/l	5.0	4.2	16
Total phosphorus, mg/l	8.5	7	18

Dissolved oxygen levels in the aerated lagoons have varied from 0.5 mg/l to 6 mg/l according to Cowlishaw.[9] During the spring of 1974, the storage lagoons were filled to a depth of 11.5 ft (3.5 m) and contained a combined volume of 6.52 billion gallons (24,700,000 m^3).[9] In the spring of 1974, a portion of the soil cement lining on the east storage lagoon dike was damaged by ice action. The water level was lowered during the summer irrigation cycle and repairs costing an estimated $90,000[13] were undertaken in the fall. In order to relieve the storage load, permission was granted by the Michigan Department of Natural Resources to discharge water from the east lagoon in June to a drainage channel leading to Mosquito Creek. The water quality was adequate to permit the direct discharge.

In May of 1974, 4500 acres (1823 ha) of corn were planted including varieties having maturities of 70 to 110 days.[9] Irrigation water was deficient in nutrients because of the long storage period. Insufficient supplies of supplemental fertilizers were ordered, which caused inadequate nitrogen availability and a yield of about one-half of what could normally be expected.[9] Despite the nutrient deficiencies and the relative lack of irrigation control, the yield was estimated at 50 bushels of corn per acre (3925 kg/ha) with a total farm yield of 225,000 bushels (7,150,000 kg). The revenue from the sale of the corn was estimated by Sheaffer[13] at $375,000. He further estimated the 1975 yield at 400,000 bushels (12,700,000 kg) of corn, producing a revenue of $1,000,000. At this level on 6,000 acres (2430 ha) the farm income will cover most of the system O and M costs.

The success of the project depends on whether the quality of discharge from the drain outlets meets the desired water quality standards because this water from the subsurface drains eventually finds its way into the surface streams in the area. Initial measurements of quality from the drain tile and the north and south outlets are shown in Table 13.4.

Table 13.4. Drainage Water Quality[a]

Parameter or Constituent	Drain Tile	North Outlet[b]	South Outlet[c]
BOD_5, mg/l	< 1	4	4
COD, mg/l	NA[d]	35	25
SS, mg/l	NA	5-20	30
Ammonia nitrogen, mg/l	0.1	0.5[e]	0.5[e]
Nitrate and nitrite nitrogen, mg/l	3[e]	0.6[e]	0.7[e]
Total Kjeldahl nitrogen, mg/l	0.1	1.4	1.2
Orthophosphate, mg/l	ND[f]-0.04	ND-0.25	ND
Total phosphorus, mg/l	NA	ND-0.2	ND
Fecal coliform, mpn/100 ml	< 1	NA	NA

[a]After Cowlishaw.[9]
[b]Tributary to Mosquito Creek
[c]Tributary to Black Creek
[d]NA means not available
[e]After urea was applied to land
[f]ND means not detectable.

The tile effluent is of better quality than the outlet flows because rubbish and the growth of aquatic plants in the ditches degrade the water. As the outlet flows increase during normal irrigation seasons, the scouring and flushing should cleanse the ditches and improve the water quality.

There were a number of reported pipeline failures[9] during pressure testing and initial startup. The pipe materials used were asbestos-cement and found to be within the contract specifications. Efforts are being made to control transient pressure surges by modifying valve operation.

CONCLUSIONS

The Muskegon project is accomplishing the philosophical points made by Boyer and Reid:[14]

> The environment is one system, with air, land, and water interacting.
> This system, for planning purposes, is closed.
> Wastes are potential resources out of place.

We can no longer think in terms of "dumping" or of concepts wasteful of energy and natural resources. The land has a tremendous capability for assimilating and using man's wastes. We must utilize this capability.

REFERENCES

1. Environmental Protection Agency. "Wastewater Treatment Project, Muskegon County, Michigan–Final Environmental Impact Statement," Publication No. PB200 348F, Washington, D.C. (1971).
2. Quarles, J. "Memo on Land Treatment," Environmental Protection Agency, Washington, D.C. (November 1, 1974).
3. Muskegon County Board and Department of Public Works. "Engineering Feasibility Demonstration Study for Muskegon County, Michigan Wastewater Treatment-Irrigation System," Federal Water Quality Administration, U.S. Department of the Interior, 11010 FMY 10/70, Muskegon, Michigan (1970).
4. Bauer, W. J. and D. E. Matschke. "Large Wastewater Irrigation Systems; Muskegon County, Michigan and Chicago Metropolitan Region," *Symposium on Land Treatment of Wastewater* (University Park, Pa.: Pennsylvania State University Press, 1972).
5. Wilson, C. D. "The Muskegon County Wastewater Management System," Fourth Environmental Engineers Conference on *Land Disposal of Municipal and Industrial Wastewaters* (Bozeman, Montana: Montana State University, 1973).
6. Chaiken, E. I., S. Poloncsik, and C. D. Wilson. "Muskegon Sprays Sewage Effluent on Land," *Civil Engr.* (May 1973).
7. Forestell, W. L., "Sewage Farming Takes Giant Step Forward," *The American City* (October 1973).
8. Postelwait, J. C. and H. J. Knudsen. "Some Experiences in Land Acquisition for a Land Disposal System for Sewage Effluent," *Conference on Recycling Municipal Sludges and Effluents on Land* (Champaign, Ill.: National Association of State Universities and Land Grant Colleges, 1973).

9. Cowlishaw, W. A. "Update on Muskegon County, Michigan, Land Treatment System," ASCE Annual Meeting and National Environmental Engineering Convention, Kansas City, Missouri (October 1974).
10. Hall, G. W. "The Development of a Center Pivot Wastewater Irrigation Machine," Paper No. 74-2554 presented at ASAE Annual Winter Meeting, Chicago, Ill. (December 1974).
11. Sprinkler Irrigation Association. "Guide to the Use of Center Pivot Irrigation Equipment for Waste Treatment of Effluent," Silver Springs, Maryland (1975).
12. Christiansen, J. E. "Irrigation by Sprinkling," Bulletin 670, University of California, College of Agriculture, Agric. Expt. Sta., Berkeley, California (1942).
13. Sheaffer, J. R. "Reuse of Wastewaters for Irrigation," Address to Sprinkler Irrigation Technical Conference, Atlanta, Georgia (February 23-25, 1975).
14. Boyer, H. and B. Reid. *Recycling on the Land–An Alternate for Water Pollution Control* (Washington, D.C.: Natural Resources Defense Council, 1973).

INDEX

INDEX